Cambridge Farm Institute Series

GENERAL EDITORS: T. B. WOOD, M.A.

E. J. RUSSELL, D.Sc.

FUNGOID AND INSECT PESTS OF THE FARM

FUNGOID AND INSECT PESTS OF THE FARM

BY

F. R. PETHERBRIDGE, M.A.

BIOLOGICAL ADVISER, SCHOOL OF AGRICULTURE, CAMBRIDGE

Cambridge:

at the University Press

1923

CAMBRIDGE
UNIVERSITY PRESS

University Printing House, Cambridge CB2 8BS, United Kingdom

Published in the United States of America by Cambridge University Press, New York

Cambridge University Press is part of the University of Cambridge.

It furthers the University's mission by disseminating knowledge in the pursuit of education, learning and research at the highest international levels of excellence.

www.cambridge.org
Information on this title: www.cambridge.org/9781107658127

© Cambridge University Press 1923

First edition 1916
Second edition 1923
First published 1923
First paperback edition 2013

A catalogue record for this publication is available from the British Library

ISBN 978-1-107-65812-7 Paperback

PREFACE

THIS book has been written for those who wish to acquire some practical knowledge of farm and garden pests. It does not aim at dealing with all the numerous diseases which affect crops but rather at giving an accurate account of some of the commoner forms. It is hoped that a working knowledge of these pests will be of great commercial value to farmers and market gardeners, and also that by their own observations they will be enabled to co-operate with those who are engaged in studying plant diseases in finding the most economical means of preventing the increasing annual losses due to these pests.

Direct cures of some of the diseases are unknown; but even in these cases an accurate knowledge of the enemy will enable us to carry out remedial measures which will prevent big losses.

Figs. 29, 31 *b*, 32, 34 *a*, 36 *a*, 37, 48, 49, and 50 are taken from Curtis's *Farm Insects*. Figs. 11, 14, 23, and 24 from *Diseases of Cultivated Plants and Trees* by Massee (Duckworth & Co.), and Figs. 4 *b* and 21 from Percival's *Agricultural Botany* (Duckworth & Co.).

My thanks are due to Prof. Biffen and the editors for suggestions and criticisms and to the above-mentioned authors and publishers for kind permission to reproduce certain illustrations.

F. R. P.

SCHOOL OF AGRICULTURE,
CAMBRIDGE.
March, 1916.

PREFACE TO THE SECOND EDITION

RECENT research has necessitated the rewriting of certain portions, and especially the accounts dealing with the Frit Fly and Wheat Bulb Fly. I wish to acknowledge my indebtedness to various workers for much of the information contained in these pages.

Those interested are reminded that they can now obtain advice on pests and diseases free of charge from the local adviser whose address will be furnished on application to the Ministry of Agriculture.

F. R. P.

January, 1923.

CONTENTS

PART I

CHAP.   PAGE

I.   INTRODUCTION TO FUNGI 1

II.   POTATO DISEASE AND ALLIED DISEASES . . 17

III. FINGER AND TOE, AND WART DISEASE . . 35

IV. MILDEWS 47

V.   ERGOT AND CLOVER SICKNESS . . . 56

VI. RUSTS 64

VII. SMUTS 73

PART II

VIII. INTRODUCTION TO INSECTS 84

IX. BUTTERFLIES AND MOTHS 98

X.   BEETLES 103

XI. FLIES 127

XII. APHIDES AND SAWFLIES 158

XIII. EELWORMS 165

APPENDIX 172

INDEX 175

PART I

CHAPTER I

INTRODUCTION TO FUNGI

It is well known that the yields of the common farm crops show enormous variations which may be due to a number of causes, e.g. kind of soil, variety of seed, management and climate. Often, however, very different yields may be obtained when the above conditions are almost identical, and in these cases it is usual to attribute the loss of crop to a "disease." Diseases of plants then may be looked upon as causes which prevent normal growth.

It is usual to classify plant diseases into two groups:

(1) Those caused by unsuitable surroundings, such as unfavourable conditions of soil, or of weather.

(2) Those caused by living agencies.

It is this second group, the diseases caused by fungi and insects, with which we are here concerned.

The two groups are not however strictly separable, for we often find plants weakened by some condition of their surroundings marked out for attack by a living agency.

The mangolds at the Rothamsted Experimental Station often suffer from leaf spot caused by a fungus known as *Uromyces betae*, but not all the plots are

equally susceptible to the disease. The plots which
have received the largest quantities of nitrogenous
manures but no potash suffer most. Except for the
manuring all the plots are under similar conditions.
Variations in food supply therefore seem to be important
in that they cause varying susceptibility to disease.

The fungi belong to the lowest group of the plant
kingdom and, as we shall see later on, cause a large
number of plant diseases. They differ from ordinary
green plants in that they do not contain any of the
green colouring matter so characteristic of the latter.
It is this substance, known to botanists as "chlorophyll,"
which enables green plants to take from the air certain
food materials that they are unable to obtain from the
soluble substances in the soil. A fungus is unable
to live in this way on the constituents of air and soil,
but requires its food to be manufactured for it, and so
it takes advantage of the food materials made by
other plants. Some live on the decaying remains of
plants and animals, e.g. toadstools live on leaf mould
and also on manure heaps; others can only take their
food from living plants, or in a few cases from animals.
Those living on dead materials are known as *Saprophytes*
and those on living plants as *Parasites*. There is no
definite boundary line between these two classes:
some members are capable of living as parasites at
one period of their life and as saprophytes at other
periods when the living food supply is not at hand.

Among cultivated plants are forms which differ very
considerably in appearance and this naturally is found
in other groups of plants. It is difficult to see any
resemblance between the mushroom which we eat and
the smut that blackens corn, or the yeast used by

brewers, or the mould that grows on jam or decaying fruit, and yet all these belong to the group of fungi. The best method of studying fungi is to grow them. The following experiment shows an easy way of doing this: allow a piece of bread to stand exposed to the air for a few hours. Then moisten it, and place it on a piece of damp blotting paper under a glass or bell

Fig. 1.   Mucor growing on bread.

jar in a warm room. In a few days the surface becomes covered with white tufts having a velvety appearance. Under the microscope these appear as a number of white interlacing threads which are part of a fungus plant and collectively known as its *mycelium*. If one of these mycelial threads known as a *hypha* is carefully examined it is seen to consist of a tube-like structure containing a somewhat granular substance which is the living portion or *protoplasm* of the fungus. Some of the tubes have no dividing walls but others are divided by cross walls into compartments called *cells*,

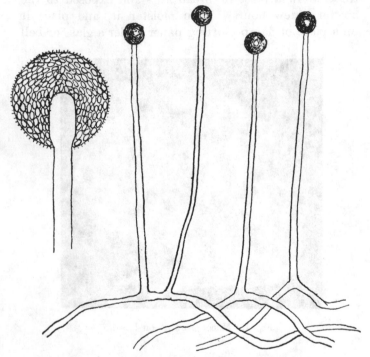

Fig. 2. Spore bearing cases of Mucor. (Magnified.)

the length of which varies with the fungus under observation. Further examination of the tufts on the bread shows that all the threads do not belong to the same fungus. Some of them carry tiny balls, at first white and later on turning black, each being borne on a long stalk (see Fig. 2). This fungus is known as a *Mucor*. Other threads give rise to a greyish powder, which under the microscope is seen to consist of numerous egg-shaped bodies like bunches of grapes. These are the seeds or, as they are usually termed, the *spores* of the fungus known as a *Botrytis* (see Fig. 3). We may also find a fungus bearing light-blue spores

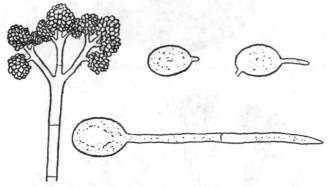

Fig. 3. Spores of Botrytis borne like bunches of grapes and some of the spores germinating. (Magnified.)

borne in chains at the ends of branches which join to form a single stalk. This fungus is a *Penicillium* (see Fig. 4).

Spores of any of these fungi placed under suitable conditions give rise to the same kind of fungus plant or mycelium as that on which they were borne. The following is a means of studying the growth of spores. Take a *Ward's Tube* (see Fig. 5) and fix the base of the chamber to a glass slide by means of paraffin wax; put in a small quantity of water. Take some of the spores of the Botrytis fungus on a brush and put them

Fig. 4 a. Spores of Penicillium as seen under the microscope.

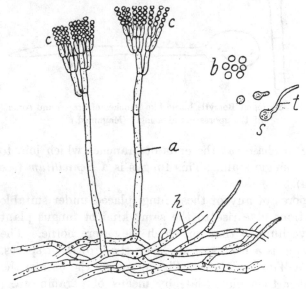

Fig. 4 b. (a) Erect hyphae bearing conidia (c) in chains. (b) Detached
conidia. (s) Germinating conidia. (t) Germ tube. (h) Mycelium.
(Magnified.) (After Percival.)

into a small drop of water on a coverslip large enough
to cover the top of the chamber; smear the top with
vaseline and place on it the inverted coverslip.   The
spores are now in a hanging drop in a moist chamber
and can be watched under the microscope.   After
about 12 to 24 hours the oval spores send out small
projections which continue to grow;   later on cross
walls are formed (see Fig. 3).

In water growth ceases as soon as the food in the
spore is used up.   If, however, a food solution is used
for a hanging drop instead of the water, the tube
continues to grow, branches and gives rise to the

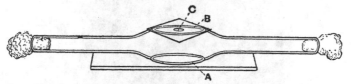

Fig. 5.   A Ward's tube.   *A*, paraffin wax; *B*, vaseline; *C*, the
hanging drop.

same kind of mycelium as that on which the spore
was borne.

A method now in common use for the study of
fungi consists in making cultures of them in such
substances as plant extracts, meat extracts, gelatin,
or agar-agar (prepared from sea-weed) or a mixture of
some of these.   The medium is freed from living
organisms by means of heat and then infected with
fungus spores or pieces of mycelium.   Saprophytes and
many parasites grow and form spores;   some parasites,
however, such as rusts, will not grow in this way.   The
experiment is usually carried out in tubes stopped with
cotton wool or in Petri dishes (see Fig. 6).

In neither case can foreign organisms enter the culture. It is ·thus possible to grow a fungus free from all other organisms. If at first the culture is not pure a few spores of the required fungus may be transferred to a sterile medium and a pure culture so obtained. The mushroom fungus resembles those already examined in consisting of a mycelium and a part which bears the spores. The spawn contains a

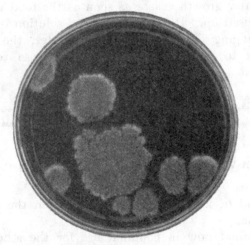

Fig. 6.   Colonies of a fungus growing on gelatin in a Petri dish.

dense growth of hyphae which take up food from decaying matter and then produce the fructification known as a mushroom.

If some of the white strands (i.e. mycelium) from mushroom spawn are placed in decaying manure or rich soil containing much organic matter, and kept moist and at a suitable temperature, mushrooms will be formed. If however the soil contains very little organic matter, no mushrooms will form. This is because the mushroom does not live like a green plant, but requires organic matter for its food.

Remove the stalk of a ripe mushroom and place it with the coloured part downwards on a piece of white paper and leave it for some time. Then tap it gently and lift it; on the paper will be found a print similar to the gills or coloured part of the mushroom which is found to consist of coloured spores (see Fig. 7). These are borne in pairs on some of the cells on the sides of the gills. The mushroom is the spore-bearing part of the fungus and it can be produced from these spores as well as from spawn. Toadstools similarly bear spores.

So far we have been growing the fungi on dead materials and dealing only with saprophytes. Parasitic fungi can be obtained by growing plants under suitable conditions for fungus attacks. Sow cress seed in a pot of soil, keep the soil very wet and in a fairly warm atmosphere. The young seedlings soon wilt and die. Examination of the surface of the soil shows a ramifying growth of mycelium, some of which has entered the plant. Here, then, we have a case of a fungus capable of injuring a plant and therefore a parasite, but it is also capable of living on the soil as a saprophyte. There are many others besides this one capable of living either as saprophytes or parasites. Some however can live only as saprophytes, e.g. Penicillium, and others only as parasites, e.g. rusts.

In a previous experiment we simply moistened a piece of bread, and on leaving it found that fungi began to grow. How did these get on the bread? The following experiment affords an answer. Take a piece

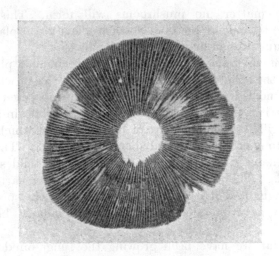

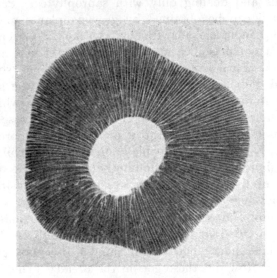

Fig. 7. Spore prints of common toadstools.

of bread as before and keep it for a short time in boiling water. Now divide the bread into two parts and place on separate pieces of glass which have also been immersed in boiling water. Cover one piece, while still hot, with a small bell-jar—call this *A*. Allow the other piece (*B*) to cool and to remain uncovered for two or three hours, and then cover it also with a bell-jar. After a few days a whitish fungus appears on *B*, but not on *A*. Now the only difference in treatment of these two pieces of bread is that *B* has been exposed to the air for a short time while *A* has not. As *A* has no fungus growing on it we may conclude that the fungus has not been produced by the bread. The only other source from which it could have come is the air. Fungus spores which we have already examined are very light and can readily float about in the air. The moist bread offers suitable conditions for the germination of any which may find their way to it; the hyphae take up food from the bread and produce a mycelium which gives rise to similar spores some of which in turn may fall on the bread and germinate as before.

Considerable use is made of the knowledge that fungus spores float about free in the air, but can be killed by heat. In jam making the jam is heated and any fungus present is killed. The jam is then tied down *whilst still hot* and any spores that reach the surface before tying down are killed, consequently there is no growth of mould on the surface. If however the pots are allowed to cool before being tied down, any fungus spores that reach the surface after the temperature has fallen too low to kill them grow and produce a mycelium which is commonly known as "mould."

In tinning fruit the tin and its contents are heated and sealed up whilst still hot; hence foreign organisms, such as fungi, are prevented from growing on them. Use is also made of the fact that fungi will not grow in certain substances known as "preservatives"; e.g. boric acid is added to cream to prevent the growth of organisms which would turn it sour.

These fungi are saprophytes, but a large number of parasites which cause plant diseases also spread themselves by means of spores floating in the air. Spores are produced in very large numbers and only by chance reach a substance on which they can feed as they have no special guidance towards it. Some give rise to several kinds of spores which were regarded as separate forms until it was found that the spores of one stage were capable of producing a mycelium bearing the spores of another stage. Every fungus is distinct from every other fungus: the spore of a mildew never gives rise to anything but a mildew, nor the spore of a rust to anything but a rust fungus, no matter on what plant it is growing, just as an acorn never gives rise to anything but an oak tree, or a grain of wheat to anything but a wheat plant.

Fungus spores can be divided into two groups:

(1) Those formed directly from the mycelium, usually by the cutting off of portions by transverse walls. These are known as *Conidia*.

(2) Those formed as the result of fertilization, i.e. the fusion of a male and female cell.

Up to the present we have only dealt with conidia; these are usually produced under favourable conditions, such as obtain in summer, and are often called "summer" spores.

Those resulting from fertilization usually serve to carry the fungus through bad conditions, such as winter. They are known as "winter" or "resting" spores, and some are capable of living for several years.

Many plants have other methods of reproduction besides seed, e.g. potatoes are usually derived from tubers, i.e. underground stems. Certain fungi have very similar methods. Portions of the mycelium of the potato disease fungus are capable of living in the potato tuber and of producing a further growth of mycelium and spores.

Mushroom spawn is capable of producing mushrooms because it contains the mycelium of the fungus.

Parasitic fungi live in much the same way as those which grow on bread. The plant which is being attacked is known as the "host." The mycelium usually lives inside its host and absorbs the food materials found there; it appears on the surface only for the purpose of forming spores. The visible portions of the fungus are the organs of reproduction and the parts of the mycelium bearing them. In the case of the true mildews, however, almost the whole of the fungus is on the outside and the food is obtained by suckers called " haustoria " penetrating into the plant.

Fungoid diseases are spread from one plant to another chiefly by means of spores which may be blown about in the air or carried by insects; some spores can live for a long time in the soil and infect plants sown there. In other cases the mycelium itself is capable of infecting plants by growing through the soil from one plant to another.

Enormous losses result from fungus attacks, but the harm done by a particular fungus varies from year

to year.   A fungus may become so abundant in certain
seasons as to cause considerable loss; such epidemics
are usually due either to very favourable conditions
for the fungus or to the introduction of a foreign
fungus.   A cucumber grower estimated his loss from
a fungus disease as £1000 in one season.   The potato
famine in Ireland in 1845 was the result of an enormous
reduction of the potato crop by the potato disease
fungus, and occurred in a season when conditions were
favourable to its growth.

The American gooseberry mildew as its name
implies came to us from America.   It first appeared in
1900 and for a few years did comparatively little
damage, so that no means were taken to destroy it
and to prevent re-importation.   It has now spread
over all the gooseberry districts in England and causes
so much damage as to threaten the existence of those
varieties which are most liable to its attacks.   If means
(however drastic) had been taken at first to destroy
all infected bushes and also to prevent re-importation,
this disease might never have established itself here.

Many countries adopt special precautions against
the introduction of diseases. ⁀ In South Africa potatoes
are imported from this country only when accompanied
by a certificate from the Board of Agriculture that
they are free from certain diseases, that these diseases
are not present in the area in which they were grown,
and that they were packed in clean new packages.
On arrival in South Africa they are examined by experts
and are not admitted if found to be diseased.   In this
way the black scab and other diseases of potatoes are
kept out of South Africa, thereby preventing considerable loss to the potato crop.

The various fungi behave very differently on different plants and also in their manner of attack. The mycelium of the potato disease fungus kills the cells with which it comes in contact and then dies; the living mycelium occurs only in and near the region of living cells.

In rusts the mycelium lives together with the protoplasm of the cells and both take their food from the same source.

The mycelium of the smut fungi lives with the young cells at the tips of the shoots without killing them. When however it enters the grain all the cells of the grain are used up and the mycelium produces spores.

In the case of "finger and toe" in turnips, the cells containing the fungus not only do not die but are stimulated to increased growth with the result that they are much larger than the surrounding cells which contain no fungus.

Some fungi live on one kind of plant only and are apparently incapable of living on others. Other fungi live only on plants belonging to a certain natural order. Others again live on a very large number of plants belonging to different orders. Rotation of crops is often very effective in keeping diseases in check, since a fungus disease of one crop is often incapable of attacking the following crop and may die out before a suitable host crop is grown. But if suitable weed hosts are present the disease may be carried on for an indefinite period. In some cases the fungus may have to travel a considerable distance before reaching a suitable host.

Our present system of marketing in which empties

from one district are often sent to other districts is very favourable to the spread of fungus diseases.

The fungus known as *Erysiphe graminis*, which causes mildew on cereals and grasses, really consists of several different fungi resembling each other in structure but differing in their behaviour towards plants.

Put some spores from a mildewed wheat plant on to a healthy wheat plant under warm and moist conditions—infection takes place. Now repeat the experiment using a healthy barley plant, and no infection of the barley results from the wheat mildew. Similarly the mildew from barley will infect barley plants but not wheat plants.

Take some spores from a mildewed wheat plant and put them on a barley leaf which has been bruised, the leaf now becomes infected with mildew.

The fungus causing potato disease attacks only plants belonging to the same order as the potato. The spores are blown about in the air and fall on a number of other plants but no infection takes place. It is not known why a fungus is capable of infecting some plants and not others.

Some fungi are confined to definite portions of the plant such as the root, the stem, or the leaf, whereas others attack the whole plant. Some cannot enter the healthy surface of a plant but can enter by means of a wound. The fungus *Nectria ditissima,* which causes "apple canker," is a wound parasite. Bad attacks of canker often follow an attack of "woolly aphis," which causes wounds through which the fungus can enter.

In order to deal successfully with a disease it is first necessary to find out its cause. If it proves to be a fungus we must then learn all we can about its life history, such as the times and methods of entering the plant, the times and methods of reproduction and the plants on which it is capable of living. From these data we may be able to find weak spots at which we can successfully get the upper hand or prevent it from spreading to other plants. Without such knowledge we may be attacking the fungus at the period when it is least vulnerable.

CHAPTER II

POTATO DISEASE AND ALLIED DISEASES

Potato disease (Phytophthora infestans).

We must not gather from the name of this disease that it is the only one from which the potato plant suffers. Potatoes are subject to a number of diseases but this one is so called because it is very widespread, occurs in all parts of the world where potatoes are grown, and because it does the greatest amount of damage. It has other names such as potato blight, late blight of potatoes, potato mildew, and dry rot of potatoes. July is the month to look for signs of the disease. The first indication of its presence is the appearance of brown patches on the leaves. These may appear on any part of the leaf, but are usually found at the tips or edges (see Fig. 8). On the under surface of these brown patches a pocket lens reveals— near the healthy part of the leaf—a number of white threads like those of the moulds on bread. Further examination under the microscope shows that the white threads consist of a number of branched hyphae carrying lemon-shaped bodies (Fig. 9 A). This is the

Fig. 8.   A potato leaf showing the patches which indicate the presence of the potato disease fungus.

fruiting stage of the fungus which causes the potato disease.

These small bodies are the spores or conidia of the fungus.   Each conidium is formed by the swelling of

the end of the branch producing it; when the swelling
has reached a certain size a transverse wall divides it
from the rest of the branch. The branch then goes on
growing and so the conidium falls off or remains at the
side. The end of the branch then swells again and
produces another conidium. In this way a branch
may produce several conidia. We shall probably find
some at the tips and others at the sides of branches.
Those at the sides are the first formed, the lowest on
the branch being the oldest. The portion of the
mycelium bearing the conidia is known as the
conidiophore and is rather peculiar. The tube which
goes on growing after the conidium is produced is
wider than the portion below, and gives a knotted
appearance due to the variation in the bore of the
hyphal tube (Fig. 9 A). The conidiophores reach the
exterior through the breathing pores of the leaf either
singly, or several through one pore. Together with
the spores they provide a means of recognizing the
fungus. The conidia may vary somewhat in size, the
average length being about $\frac{1}{50}$ in. and the average
width being about $\frac{1}{130}$ in.

The conidia are very light and readily blown about,
and as the plants overlap and touch each other they
may easily fall or be washed by the rain from one plant
to another. It is difficult to watch the development
of these conidia after they have fallen on a potato
leaf, but the further development may be followed
by putting some of them in a hanging drop in a
suitable chamber (see p. 5). Here changes take place
similar to those which occur naturally when the
conidia fall on moist leaves. After a time a change
occurs in the contents of the conidium, there being a

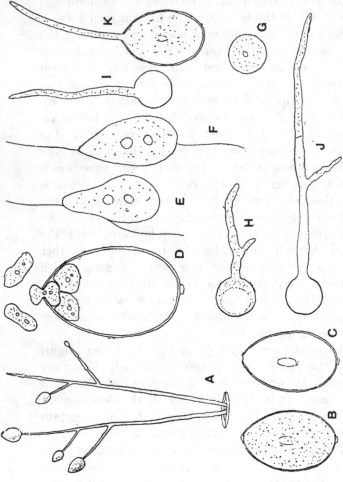

Fig. 9. *Phytophthora infestans* (the potato disease fungus). *A*, conidiophores bearing conidia; *B*, a conidium; *C*, an empty conidium; *D*, zoospores escaping from a conidium; *E* and *F*, zoospores; *G*, a zoospore just before germinating; *H, I* and *J*, germinating zoospores; *K*, a conidium germinating directly. (Magnified.)

suspicion of a division, but not a very marked one. Then suddenly the whole of its contents come pouring out through a small hole at the pointed end. The whole process may take only about 10 seconds. The contents do not pass out whole but in about ten separate masses. Each mass is considerably larger than the hole through which it emerges and consequently is much constricted on passing out (see Fig. 9 *D*). On arriving outside the conidium these masses start swimming about very rapidly. It is from this motile power in which they resemble small animals that they are called *Zoospores*. When one of them becomes motionless the presence of two small threads is revealed (Fig. 9 *E* and *F*). By means of these the zoospore moves; they lash the water and force it along. After a time the zoospore ceases its movements and the threads disappear; it now loses its former shape and becomes spherical (Fig. 9 *G*). Later on a small protuberance is formed; this lengthens and forms a tube. The contents of the zoospore pass into this tube which grows and may give off branches (Fig. 9 *H*, *I* and *J*). Eventually growth stops and the young fungus dies of starvation.

All these changes take place when a conidium falls on the surface of a moist potato leaf, but the hyphal tube formed does not die of starvation; it enters the leaf by means of a breathing pore or by dissolving away the outer wall of the leaf. Growth now goes on rapidly as the hypha has abundant food. A mycelium forms which kills the leaf cells and changes them to a brown colour causing the characteristic appearance which first shows the presence of the disease. Both young and old leaves may become infected in this way.

Every potato grower knows that the disease is very bad in some years which become known as potato disease years and much less harmful in others. The disease is bad when dull, muggy weather prevails in July and August. Farmers formerly held the idea that the dull weather caused it; indirectly it does, for damp conditions are favourable to the growth and spread of the fungus which produces the disease. Potatoes grown in a shady place with a moist atmosphere readily succumb to the disease, whereas others in a dry sunny place may show no signs of it. We are now in a position to understand why the disease should be most marked in dull weather. The conidia and zoospores by which it is distributed can only germinate when a certain amount of water is present and at certain temperatures. The optimum temperature has been found to be about $70°$–$74°$ F., and this is about the temperature of dull weather in July.

If the weather is bright and hot the brown patches formed on the leaves dry up and curl, and the disease may not spread at all or only very slowly. If, however, the weather is dull the brown patches rapidly spread over the whole leaf. The formation of conidiophores on the under surface accompanies the browning of the tissues in its spread over the leaves. It is at this stage that the casual observer usually notices the disease. Brown patches also appear on the stem and in bad attacks all the leaves are killed. A diseased field has a very bare appearance, similar to that of a healthy field when the leaves have died normally.

The tubers of diseased plants are marked in a characteristic manner. They have sunken brownish patches which at first do not extend very deep into the tuber. Later on however the browning may spread throughout the tuber. These patches may show themselves soon after the leaves have become diseased, especially on tubers near the surface. This stage of the fungus has given it the name of "dry rot."

Tubers which are apparently healthy have been sown, and the plants from them grown under conditions rendering external infection impossible, and still these plants have become diseased. Careful examination of a number of these apparently healthy tubers showed them to contain portions of the mycelium of the potato disease fungus. The disease then starts from the tuber. When the tuber sprouts and produces shoots the fungus mycelium also grows and makes its way into the stems and leaves. We have seen how the disease reaches the leaves and spreads from one plant to another, but we have still to account for the infection of the new tubers which carry the fungus on to the following season.

We should expect to find that the fungus mycelium grows down the stem and into the new tubers. The evidence on this point is not very clear. It is supposed that the new tubers are capable of being infected from the conidia produced on the leaves. These conidia fall on the damp earth and there produce zoospores which are washed down to the tubers. The zoospores germinate and penetrate the tuber, the mycelium formed causing the brown patches so characteristic of the disease. In favour of this hypothesis are the facts (1) that the diseased patches appear as frequently at the end of the tuber away from the stem as at the end which is attached to the stem; (2) that the upper surface is often more diseased than the lower; (3) that

tubers near the surface are often more diseased; and
(4) that the browning of the tissues extends inwards.
It has been shown that zoospores will not pass through
a depth of 6 inches of sandy soil. An experiment was
made by covering potatoes with this depth of soil to
see if the tubers were thus prevented from becoming
diseased. Although a number of the tubers became
diseased the percentage was much less than in the
case of tubers covered with a lesser depth of soil. It is
important to know how the young tubers are infected,
because if it is entirely by means of the zoospores we
should hope to aim at preventing their infection by
trying to keep down the formation of conidia or to
prevent the zoospores from reaching the young tubers.
It is possible that the tubers are infected directly by
the growth of mycelium through the stem as well as
indirectly by means of the zoospores.

By killing the leaves of the potato plant the disease
stops their functions as feeding and breathing organs,
and as the plant is now capable of taking only very
little food from the air the growth of the tubers practically ceases. Having lessened the crop and reduced
the value of the tubers lifted the disease has by no
means finished its career. When slightly diseased
tubers are stored in clamps, the disease spreads,
especially when there is a sufficient supply of moisture
and a suitable temperature. Diseased tubers also
allow foreign organisms to enter thereby setting up the
well-known "wet rot."

Take some diseased tubers, wash them, cut slices
across the diseased areas with a sterilized knife and
place them in a large sterilized Petri dish lined with
moistened filter papers.

After a time white tufts will grow on the slices of potato. Examine these and they will sometimes be found to consist of the conidiophores of the potato disease fungus, showing that the mycelium of that fungus is present in the tubers. Sometimes the white tufts are portions of saprophytes which have gained an entry into the tuber, and when present these often make it difficult to find the potato disease fungus.

Remedial Measures.

The life history of the fungus causing potato disease provides us with a sound basis for deciding what steps to take in order to minimize the amount of damage done. Moisture is one of the conditions favourable to the disease, and so, if possible, in choosing a soil in which to plant potatoes preference should be given to one that is well drained. Having chosen our soil we must then exercise some care in choosing our seed. The disease is carried on from one season to another by means of mycelium in the tubers. Practically all the potatoes from badly diseased fields contain this mycelium, although on many there may be no outward sign of it: consequently we shall be helping the disease if we plant them. Seed should be chosen from fields with the smallest amount of disease.

In choosing between the different varieties the grower has to consider the value of the crops produced. Those susceptible to disease are usually the high yielders and it is for the grower to decide which are the more profitable on his particular soil. In good potato years the amount of disease on the susceptible varieties is usually small and even in bad years it is possible to prevent it to a certain extent, as we shall see later.

It spreads in the field by means of conidia which form zoospores. If we can prevent the formation of spores or their germination we shall put a check on the spread of the disease. A large number of substances in the form of liquids and powders have been put on potato plants with a view to checking the damage done by this fungus, and some of them have proved successful. An application of this kind to check the spread of a fungus is known as a "fungicide." Those proved valuable in the case of potato disease are "Bordeaux mixture" and "Burgundy mixture."

Bordeaux mixture is made from copper sulphate (blue stone), lime and water. Ordinary agricultural blue stone must not be used as it does not contain sufficient copper sulphate. The blue stone used must contain 97–98 per cent. of copper sulphate. The lime most suitable for this purpose is "fat" lime, or as it is often called, "white lime." This is the kind used by plasterers. It is important that the lime should be freshly burnt and in lumps. Soft water makes a better mixture than hard water.

For making 50 gallons of the mixture the following are required:

    A tub (of about 60 gallons capacity),
    A tub (of about 30 gallons capacity),
    A wooden pail,
    5 lbs. of copper sulphate,
    2½ lbs. of lime,
    50 gallons of water.

Put 25 gallons of water into the large tub and weigh out 5 lbs. of the copper sulphate (this should be in a

powdered state). Tie the copper sulphate in a cotton
bag or a piece of sacking, and suspend it so that it
reaches just below the surface of the water in the tub.
Now carefully pick over the lime, throwing out any
pieces that are overburnt, and weigh out 2½ lbs. of it.
Put this into the wooden pail and slake it with about
a quart of water which should be added very slowly.
Covering the bucket with a piece of sacking helps the
lime to slake.   When the lime has crumbled add about
another quart of water and work the lime into a stiff
paste; add more water and stir to make a creamy
liquid. Now put 20 gallons of water into the small
tub, pour the creamy liquid into it through a fine
sieve or a piece of coarse sacking, and add water until
the tub contains 25 gallons.

When the blue stone in the large tub has all dissolved slowly pour the lime solution into it, stir well
meanwhile and continue to do so for some time. If
the above operations have been carefully carried out
the mixture should consist of a light blue precipitate
suspended in a colourless solution. A granular precipitate which sinks quickly to the bottom shows the
mixture to be a poor one. It should remain floating
for some time. In some cases Bordeaux mixture
damages the potato plant. This is usually due to
some error in making and is caused by the presence
of free copper sulphate. It is a simple matter to test
the safety of the mixture. If a knife blade becomes
coated with copper when placed in the liquid, it is
unsafe and more lime solution should be added.

A better test is to add a few drops of the mixture
to a solution of potassium ferrocyanide on a piece of
white porcelain. If a brownish colour is produced

more lime solution must be added. If the above directions are carefully followed the mixture will usually be satisfactory.

The mixture is sprayed on the potato crop by means of a pumping apparatus to which a special nozzle is attached. For large areas a horse-drawn machine is necessary. For small areas the potatoes can be sprayed successfully by means of a "knapsack" sprayer which is carried on the back.

The object of spraying the plant is not to cure the disease but to prevent it from spreading by covering all the parts liable to infection with a coating of a substance which will prevent the spores from germinating, or kill the delicate germ tube, and so keep it from penetrating the plant. It is no easy matter to cover the plant uniformly in the desired manner. The best results are obtained by means of a very fine spray, which on coming out of the nozzle hangs about in the air for a short time like a mist. Such a spray can only be obtained by means of suitable nozzles of which there are several useful forms on the market at the present time. A much finer spray is formed when a high pressure is obtained.

Bordeaux mixture should be kept stirred while it is being applied to the plant and this should be brought about by an automatic stirrer inside the machine. With a coarse spray the mixture easily runs off the plants.

The amount of spray required is from 100–150 gallons per acre varying with the amount of haulm. The time to spray is just before the disease appears and this varies in different districts (see Appendix). In a bad season it is advisable to spray early, and to repeat about three weeks later. Further applications are desirable if the unfavourable conditions continue. The result of spraying is usually a considerable increase per acre as the plants remain green longer and consequently the tubers grow longer. This delays the lifting which in some cases may be a disadvantage. In dry seasons when the disease does little damage spraying may not increase the yield, but as the weather is so uncertain it is an insurance to the farmer that he will benefit considerably if the disease does appear. Early varieties are often dug before the disease appears and the advantage of spraying is not so marked with second earlies as with maincrop potatoes.

Burgundy mixture is often used for spraying potatoes because it is almost as good as Bordeaux mixture and is much easier to make. It consists of 4 lbs. of copper sulphate and 5 lbs. of pure washing soda dissolved in 40 gallons of water. The copper sulphate and the crushed soda are dissolved separately and then stirred into the 40 gallons of water.

There is now a Burgundy mixture in which the copper sulphate and washing soda are mixed together in the form of a fine powder.

Dry spraying is the application of the fungicide in the form of a very fine powder or dust. I have never seen as good results from dry spraying as from wet spraying but it is useful to those growers who have difficulty or expense in obtaining water.

It is economical of labour and easy to apply with a horse-drawn machine, but it can only be satisfactorily applied when the dew is on the leaves.

Wet spraying requires a certain amount of care in the making of the mixtures and in the hands of a care-

less workman harm can be done by the scorching of
the haulm if the correct weights are not used.

In hot sunny weather it is not safe to spray potatoes
if they are attacked by "green fly" (aphides), as much
damage may be caused by scorching of the foliage.

The storage of potatoes requires considerable care.
Whenever possible they should be allowed to dry before
they are covered in the "pits" or "clamps." The sound
tubers should be separated from the diseased tubers
preferably at the time of digging. If the potatoes are
wet and dirty so as to make this impossible a temporary
clamp should be made; later on the healthy potatoes
should be picked out and put into a permanent clamp
and the diseased potatoes may be fed to stock.

The site of the clamp should be as dry as possible.
It should be narrow and have steep sides in order to
let the moisture escape easily. It should be covered
with a good straight straw or similar material to keep
out the rain. It is a common practice to cover the
straw with a layer of earth. In the clamp the potatoes
heat and sweat so that moist warm conditions suitable
to the spread of the disease are obtained if it is not
properly ventilated. Other organisms also enter
bruised and diseased potatoes and set up the well-
known "wet rot."

For the first few weeks the clamp should not be
covered with earth but later on this is necessary to
keep out sharp frosts. On covering with earth spaces
should be left at the top at frequent intervals which
are covered with straw only. The tubers should be examined from time to time to see if they are keeping well.
If the potatoes are wet and dirty on lifting it is advisable
to sprinkle them with some powder to prevent the

rot from spreading in the clamps.   Ground lime at the
rate of 1 cwt. to every ton of potatoes, or flowers of
sulphur at the rate of 2 lbs. to every ton of potatoes,
have proved very successful in preventing the spread
of winter rot.

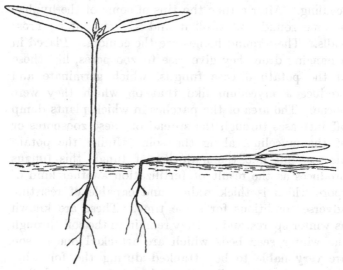

Fig. 10.   Tomato seedlings, one healthy and the other killed by
the damping off fungus *Pythium de Baryanum*.

Damping Off (*Pythium de Baryanum*).

In seed beds a number of seedlings are often found
lying prone on the ground, the stem of the plant being
rotten just at the ground level (see Fig. 10).   This
state of affairs is known as *damping off*.   There are
several fungi which cause young plants to damp off in
this way, one of the commonest being *Pythium de*

Baryanum. This fungus is capable of living for a time
on the decaying matter in soils; on reaching a plant
its mycelium bores into the cells and soon kills them.
The stem is thus weakened at the ground level and
topples over. The non-septate[1] mycelium of the
fungus may be found between the cells of a diseased
seedling. After a time the tips of some of the hyphae
become round and swollen and are cut off by cross
walls. These round bodies are the conidia. Placed in
a hanging drop they give rise to zoospores, like those
of the potato disease fungus, which germinate and
produce a mycelium like that on which they were
borne. The area of the patches in which plants damp
off increases through the spread of these zoospores or
of the mycelium along the soil. Unlike the potato
disease, but like the majority of fungi, this fungus
produces as the result of fertilization another kind of
spore which is thick walled and capable of resisting
adverse conditions for some time. These are known
as winter spores, and as they remain in the soil through
the winter, seed beds which are attacked one season
are very liable to be attacked during the following
season unless the soil is treated and the spores killed.

Remedial Measures[2].

The conditions which favour the fungus are warmth,
moisture, and shade. If possible we must try and get
rid of these by avoiding shady spots for seed beds, and
also by using soil which allows water to pass through
freely.

[1] In a non-septate mycelium the hyphae have no cross walls.
[2] See Appendix B.

In seed boxes the soil may be freed from this fungus by heating it to 150° F. Mixing charcoal with the upper layer of the seed bed reduces the disease. Take two boxes of soil in which plants have damped off. Heat one of them to a temperature of at least 150° F. Sow each of the boxes very thickly with cress seed. Keep the soil very wet and use water which has been previously heated to 150° F. Keep the boxes in a shady place. When the plants grow many of those in the unheated box will damp off, whereas those in the heated box will not. Now introduce some of the soil and dying plants from the unheated box into the heated box; the plants begin to damp off.

Onion mildew (*Peronospora schleideni*).

This disease is usually present wherever onions are grown. It is easily recognized by the yellowing of the leaves and also by the "necking" of the plant, i.e. elongation of the stem between the leaves and the bulbs. The leaves which are attacked soon collapse. Like the potato disease it does most damage in dull warm weather. On the leaves of a diseased plant a dull violet-coloured growth forms. This is the fruiting stage of the fungus and consists of tree-like conidiophores bearing conidia on the tips of their branches. These somewhat resemble those of the potato but each conidiophore bears a larger number (see Fig. 11). In water these conidia give rise to zoospores which are capable of further infection. If damp weather continues the disease spreads rapidly. In dry weather, however, it soon disappears.

This fungus also forms winter spores like those of *Pythium de Baryanum* which serve to carry the disease from one season to another. The fungus does not attack the bulbs but prevents the leaves from carrying on their normal functions of manufacturing food stuff and so the bulbs are smaller.

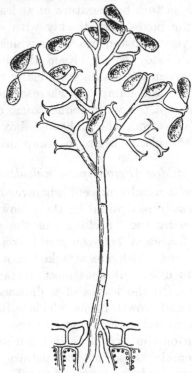

Fig. 11. Conidiophore and conidia of *Peronospora schleideni* which causes onion mildew. (Magnified.) (After Massee.)

Remedial Measures.

Well-drained land does not suffer so much as wet land. Onions should not be planted on land which had a diseased crop of onions in the previous season.

Spraying with Bordeaux or Burgundy mixtures or with Liver of Sulphur, one ounce to two gallons of water will check the disease as by this means the spores are prevented from sending their germ tubes into the plant.

CHAPTER III

FINGER AND TOE, AND WART DISEASE

Finger and Toe (Plasmodiophora brassicae).

This disease also goes by the name of "anbury," "club root," or "club." It attacks all cultivated plants of the cabbage order such as turnip, cabbage, rape, kohl rabi, mustard, stocks and wallflowers, and also the weeds, e.g. charlock, shepherd's purse, and hedge mustard. As far as we know it is confined entirely to the Cruciferous order[1] to which these plants belong. This disease is very widespread and has done a considerable amount of damage, especially since the practice of using acid manures became common and the use of chalk and lime declined.

On taking up young cabbage plants from seed beds for the purpose of transplanting, the roots are sometimes found to be swollen and knotted. These knots, which vary considerably in size and shape, may occur on any part of the root system. If these plants are put out into the field they do not grow normally; a few small leaves may be formed, but they soon

---

[1] Order Cruciferae. It is exceedingly useful in this and several other diseases to know the characteristics of this order, which are very simple and can be found in any book on Systematic Botany.

Fig. 12 a. Swede turnips attacked by Finger and Toe.

Fig. 12 b. Charlock attacked by Finger and Toe.

become yellowish and the plant makes very little growth. A diseased plant possesses, instead of the normal development of fibrous root, a swollen nodular mass. The fancied resemblance of this mass to fingers and toes gives the disease its popular name. Examination of a thin section of one of these nodules mounted in water shows large cells scattered amongst cells of

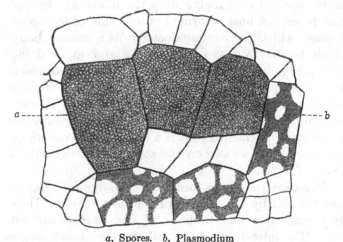

a, Spores. *b*, Plasmodium

Fig. 13.   A section of a turnip root very highly magnified showing the giant cells caused by *Plasmodiophora brassicae*. Some of the giant cells contain "plasmodium" and other spores. (Magnified.)

ordinary size. The contents of the larger ones are darker in colour and very different from those of an ordinary normal root cell. The high power of the microscope shows that the contents consist either of a frothy granular mass or of a very large number of tiny spheres (see Fig. 13). These are the spores of the Finger and Toe fungus. They are very small and measure only $\frac{1}{8000}$ of an inch in diameter.

[N.B.   Do not confuse these spores with the starch grains of the cells.   The latter are larger, irregular, and not so numerous.]

Some of these minute spores may be found floating freely in the water in which the section is mounted. Owing to their minute size their germination is difficult to watch, except under a very high power of the microscope.   In a hanging drop the following changes take place.   A hole is formed on one side of the spore through which the contents emerge as a minute body which begins to move very slowly and in so doing continually changes its shape.   Each one is known as a *Myxamoeba,* from its resemblance to a small animal which moves very similarly, called *Amoeba.*   It consists of a naked mass of living substance known as protoplasm.   Attached to it is a hair-like process which resembles those we saw on the zoospores of the potato disease fungus.

In nature the spores formed in the large cells escape into the soil by the rotting of the host plants.   Here they germinate and millions of *myxamoebae* are set free.   The interstices of the soil are very much larger than these and so there is plenty of room for them to move about.   The amount of water they require to move in is also small and there is sufficient in an ordinary moist soil.   In their wanderings through the soil water some of them may chance to reach a rootlet of a cabbage plant or a plant belonging to the order Cruciferae.   Of the large number of *myxamoebae* set free only a very small percentage probably reach a plant which they are capable of infecting.   The others of course die.   We see now why such an enormous number of spores are formed.   If only a small number were formed the chances of infecting plants would be proportionately diminished and consequently the existence of the fungus would be endangered. The *myxamoebae* make their way into the plant through the fine hairs near the tips of the rootlets.

The fungus causing Finger and Toe belongs to a group which is very different from the majority of fungi. It is not always classified as a fungus, but for our purpose may be considered as such. If these spores belonged to a fungus similar to those already studied, we should expect them to give rise to a number of tube-like hyphae known as a mycelium. In this case, however, nothing of the kind takes place; when the *myxamoeba* gets inside a root cell it begins to feed on the contents. This particular cell seems to make a call on the food supplies of the plant and gradually gets larger. The *myxamoeba* continues to feed on the cell protoplasm until eventually this is all used up. The large or giant cell is thus filled by a naked mass of protoplasm derived from the *myxamoeba* and which is now called a "plasmodium." This is the vegetative portion of the Finger and Toe fungus, and so corresponds to the mycelium of other fungi. As already mentioned it appears as a frothy granular mass.

The plasmodium is able to pass from the giant cell, which it has caused to be formed, into neighbouring cells. Here it feeds on the protoplasm and these cells gradually enlarge, and eventually become filled by it. Examination of a section shows that the giant cells touch each other and this is what we should expect from the method of infection by creeping from cell to cell. Now let us look at the case from the point of view of the plant attacked.

The plant has to give up some of its food to the development of the plasmodium, at the same time it continues to grow and manufacture fresh food material. As the plasmodium grows bigger it requires considerably more food material and consequently the plant growth is lessened until a time is reached when the plant ceases to grow, or grows only very slowly, the whole of the food manufactured by the leaves or brought up by the roots going to the support of the plasmodium.

The formation of giant cells causes the roots to become swollen, and when a portion of the fungus proceeds in a certain direction nodules are formed, until eventually a state of affairs is reached as shown in Fig. 12.

After a time the plasmodium becomes divided up into a large number of small portions, each of which becomes spherical and surrounds itself with a definite cell wall. These spherical bodies are the spores. When the host rots these spores are set free in the soil ready to start on their destructive career, when the favourable conditions of spring bring about their germination. It is supposed that they can retain their vitality for several years, which characteristic is extremely favourable to the fungus in the ordinary rotation of crops. This longevity of the spores, together with the presence of certain cruciferous weeds which can act as hosts of the fungus, enables it to continue its existence when cruciferous plants are grown every four or five years.

There are other diseases known as "club," which cause similar distortions to plants. One of these is caused by a weevil which may be found inside the swollen parts (see p. 122). Another is caused by an eelworm (see p. 170).

Finger and Toe may be distinguished from these by the characteristic "giant" cells.

In addition to its dwarfing influence this fungus causes turnips and swedes to rot quickly.

Remedial Measures.

No method is known by which this disease can be cured once it has established itself in a crop. It is possible however to carry out certain practices which may considerably reduce the loss. The disease sometimes makes its appearance in cabbage seed beds, and on pulling young cabbages to transplant they should be destroyed if attacked and others obtained for planting in the field. Considerable harm may be done by transplanting infested cabbages, for a field may become infected by putting in diseased plants. Special care is needed to prevent the disease from spreading from field to field, as it may be carried about in portions of diseased roots or in soil containing the spores. This disease is not carried through the air as are the spores of the potato disease fungus. In ordinary practice diseased crops would be folded with sheep or the roots carried to the homestead or pastures and fed there.

If the diseased roots are fed at the homestead portions of the plants will become mixed with the dung. The farmer must bear in mind that this dung will spread the disease; it should not be put on land which is to grow a cruciferous crop in the next five or six years, but used for such crops as mangolds, potatoes, carrots, parsnips, beans, cereals, or grass, as these crops are not attacked. In such cases the disease will remain in the soil for four or five years, and if a cruciferous crop such as turnip is grown it is liable to attack.

If it is necessary to use the dung for a cruciferous crop the roots should be boiled before they are fed. Diseased roots should not be put into the compost heap if the compost is to be used for the cruciferous crops. The disease very often appears on a small portion of a field and by taking measures at this stage it is easy to keep it in check. Sheep folded on such a field spread it over the whole unless the diseased portion happens to be the last portion folded. If such a patch is found it should be isolated from the rest of the field by means of hurdles, and the roots pulled and carted and treated as infected material. Care must be taken not to infect the whole field from the diseased portion, which should be limed as shown below. The disease may be spread by means of soil clinging to the feet of sheep, horses or men, and also by cartwheels and implements. Dirt should be scraped from these before they are used on other fields. The spores of *Plasmodiophora brassicae* are said to live for about five years. Consequently the rotation should be so arranged as to keep cruciferous crops, such as turnips, cabbage, rape, kohl rabi, and mustard out for at least six years. In the ordinary four course rotation turnips may be replaced by mangolds, potatoes, parsnips, etc., by which means turnips will not be taken until eight years after a diseased crop. Any crops other than cruciferous ones may be grown. It is extremely important to keep the land free from the weeds of this order for they carry on the disease from year to year. Charlock is a great offender in this respect. By growing immune crops the fungus cannot feed and so eventually it is starved. It has been found that it flourishes in sour or acid soils and not in soils containing a sufficient supply of lime. It is practically unknown on ordinary soils containing a high percentage of lime. This provides us with the key for solving the question as to why certain artificial manures favour the disease. Experiments have shown that Finger and Toe may become very bad on certain soils when acid manures are used in large quantities. Such manures as superphosphate and sulphate of ammonia use up the free lime in the soil and tend to make it sour. On soils liable to the disease through lack of lime these manures should be given up in favour of basic slag or bone meal and sodium nitrate, unless large quantities of lime or chalk are also used. It has also been shown that liming the soil keeps the disease in check. As soon as a diseased crop is raised 2 tons per acre of freshly slaked lime should be well spread over the surface. It is essential that the lime should be finely divided. If chalk is used about 4 tons per acre should be put on but it is not so effective. A similar dressing should be given about 18 months or six months before the next cruciferous crop is put on the soil. The object of putting on the lime is to get rid of the conditions favouring the growth of the fungus causing the disease. The lime also kills the spores with which it comes into contact. White turnips succumb more readily to this disease than swedes. It is important that the soil used as seed beds for cabbage should contain sufficient lime. In market gardens on light soil care must be taken not to add too much lime or chalk otherwise potatoes grown in the limed soil will suffer badly from common scab (see Fig. 15).

Wart Disease (Black Scab) of Potatoes (Synchytrium endobioticum).

This disease, which also goes by the name of cauliflower disease, potato tumour, and canker, was not known in England until the end of the last century.

Its appearance varies with the virulence of the attack. In slight attacks it may not be noticeable or may appear as small blackish warts near the eyes. In bad attacks large cauliflower-like outgrowths are formed on the shoots arising from the tuber. These are at first the same colour as the potato, but eventually become almost black. The outgrowths always make their appearance at the eyes, but later they may spread over the tuber. They vary in size and at times may be larger than the tuber on which they are borne. By taking sections of the outgrowth many of the cells may be seen to contain a number of roundish spore-like bodies with thick walls. Some of these are said to germinate directly but others do not germinate for several years. On germination one of these spore-like bodies produces a number of zoospores, each carrying one flagellum. By means of these zoospores other potato plants are infected. When a potato rots they are liberated into the soil, and potatoes planted are liable to infection. They are said to be capable of living for a number of years. The disease also spreads from slightly diseased tubers if they are planted the following season. It does not attack other crops but can infect the weeds Black Nightshade and Woody Nightshade. It is common in the west and north,

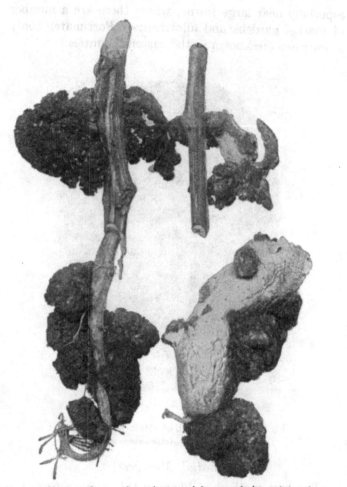

Fig. 14. Outgrowths on the tubers and leaves of the potato plant caused by *Synchytrium endobiotica*. (After Massee.)

especially near large towns, where there are a number of market gardens and allotments. Fortunately only a few cases are known in the eastern counties.

Fig. 15. Common scab of potatoes—caused by
the minute fungus *Actinomyces scabies*.

Remedial Measures.

The best method of keeping this disease from the potato crop is to exercise great care in obtaining uninfected seed. It is advisable to get seed from districts which are known not to be infected, or to have been free from infection for the last six years. Some of our good varieties such as Great Scot, Arran Comrade, Ally, Majestic, Kerr's Pink and Golden Wonder, are immune from the disease. All badly diseased tubers and other parts of the plants should be destroyed. Slightly diseased tubers may be fed, but they should be boiled beforehand, as otherwise the spores pass through the animal unharmed, and so infect the manure. This should not be put on potato land.

It should be remembered that the disease may be spread by unboiled potato peelings, on the boots of persons walking over the field and by horses, implements and soil clinging to crops grown on infected land.

Experiments with fungicides have not proved a success.

The only successful method of keeping the disease in check is to grow immune varieties.

It should be noted that this disease is scheduled under the Destructive Insects and Pests Acts and all cases must be reported to the Board of Agriculture and Fisheries.

CHAPTER IV

MILDEWS

The term "mildew" is loosely used, and often applied to a very large number of the smaller fungi, both parasitic and saprophytic It is now usual to confine the term mildew to two families. The members of the family to which the potato disease belongs are known as *Downy Mildews*. The fungi we are considering in the present chapter are spoken of as *Powdery* or *True Mildews*.

The true mildews comprise a large number of different species. One of the commonest and also one which will serve as an example of the group is the well-known mildew of grasses and cereals, *Erysiphe graminis*. We shall have little difficulty in finding specimens of this disease in cornfields during the

Fig. 16. Wheat plant attacked by *Erysiphe graminis*.

spring, especially on the lower leaves and leaf sheaths. In cases of bad attacks the leaves yellow and die. The leaves of a mildewed plant are covered with a white, a dirty white, or a pinkish powdery substance. It was to this that the term mildew was first applied, the leaves having the appearance of being dusted with some kind of powder (see Fig. 16).

In the first stages of the disease the leaves are covered with what looks like a dense spider's web. On examination this is found to consist of a network of hyphae or mycelium ramifying on the surface of the leaves and very similar to the mycelium obtained on bread cultures. The hyphae are divided by cross walls and so consist of long chains of cells. They branch irregularly and cross each other to form the dense network. Sections of the mildewed leaf show that the mycelium is not present inside (a piece may be dragged across the section on cutting and appear to be running in the interior of the leaf). It lies very close to the surface and is fixed to the leaf at certain spots; sections show how this fixture is brought about. At intervals along the mycelium short swollen tubes are given off and penetrate the surface layer. These swollen tubes are known as haustoria. This fungus differs from those already studied in having its mycelium entirely outside the plant which it is attacking. In the case of the other parasites the mycelium is entirely inside the plant and the only portions visible are the fructifications. This feature of having the mycelium outside the plant is characteristic of the "true mildews." The haustoria which it sends into the plant serve as a means of attachment and also of absorbing the food supply. In favourable weather the mycelium spreads rapidly over the whole leaf. After a time a powdery substance is produced. This powder consists of the conidia. The hyphae which produce them differ from the ordinary mycelium in growing out at right angles to the surface of the leaf instead of parallel to it. These hyphae are divided up by cross walls and the cells at the end of each become rounded

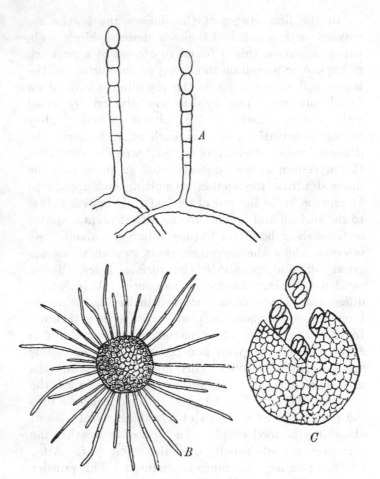

Fig. 17. *Erysiphe graminis.* *A*, conidiophores bearing conidia; *B*, a spore case with appendages; *C*, a spore case which has burst showing the asci and ascospores. (Magnified.)

off as conidia (see Fig. 17). Each cell of the hypha in its turn becomes the terminal one and forms a conidium. The formation of these gives rise to the powdery appearance. External conditions play a considerable part in their formation. In muggy weather they are produced in enormous numbers. Dampness is also required for their germination. This is why the disease is so prevalent under damp conditions. In bright sunny weather fewer erect hyphae and consequently fewer conidia are formed and conditions are less favourable to their germination. The conidia are oval in shape and collectively may be whitish grey or pink. On germinating they send out tubes which, if they happen to fall on the same kind of plant as the one on which they were formed, immediately send out haustoria into the host plant. The mycelium so produced resembles the parent mycelium. In this way the disease is spread rapidly from plant to plant.

As the season advances the mycelium has a darker appearance, and here and there among the hyphae a number of dark bodies are formed. Under the microscope these appear as a number of dark brown spherical bodies attached to the mycelium on their under surface (see Fig. 18). By separating them from the mycelium they are seen to have a shiny appearance and polygonal markings covering their surface (Fig. 17). They are the "spore cases" of the fungus, and as they get older they become darker in colour. When ripe their walls are brittle. Put some on a slide under a coverslip and tap it; the spore cases will burst open, revealing a number of somewhat pear-shaped bodies which contain a varying number of spores inside them. The latter are known as ascospores and the pear-shaped body containing them is known as an ascus. The wall of the spore case is composed of a number of polygonal cells, whose walls cause the polygonal markings on the outside. The spore cases are formed at the points where the hyphae touch one another, and each is the result of fertilization.

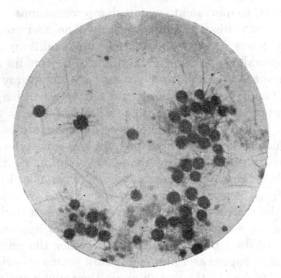

Fig. 18. Mycelium and spore cases of *Erysiphe graminis*. (Magnified.)

Attached to them are a number of "appendages." In the case of *Erysiphe graminis* these are simple, un-branched, short hyphae, but in other mildews they are longer, often branched, and serve as a useful means of distinguishing between the different mildews. The number of asci in each case is also helpful in this way. The ascospores do not germinate immediately but remain throughout the winter in their cases on the dead leaves or in the soil.

They are protected from injury due to climatic conditions by the walls of the spore case. In spring the cases burst and the asci are liberated. The walls of these also split and the ascospores are set free, and under suitable conditions germinate, sending out a short tube capable of infecting growing host plants. They germinate most readily at a temperature of about 50° F. The mycelium which they produce buds off conidia during the spring and summer. The damage done by this mildew to cereals in a wet season may be considerable, arising not only from the food which the plant gives to the fungus, but also from the limiting of the work done by the leaves in manufacturing food from the air. The normal amount of light is prevented from falling on the leaf by the presence of the mycelium. The amount of air which ordinarily enters the pores of the leaf is also reduced. The diseased leaves become yellow, and die before the normal ones, and hence their functions cease earlier, with the result that the total yield is lower.

The fungus known as *Erysiphe graminis* comprises a number of different varieties; and each one may be regarded as being distinct from the others, although no difference has yet been found in their structural appearance. The different varieties are distinguished by the plants which they attack. One variety attacks only wheat, another barley and sometimes wheat to a very slight extent, another attacks oats and a related grass. A cereal following another cereal crop which has been badly attacked by mildew will suffer no more than if it had followed a clean crop, as these varieties are so specialized that the one attacking any particular cereal will not attack the others.

Remedial Measures.

The whole of this fungus is on the outside of the plant, and consequently we should expect to be able to keep it in check by the use of a suitable fungicide. In the case of cereals the difficulties of spraying are so great that it is not usual to make application of this kind. Damp shady places should be avoided if possible as they provide conditions favourable to the growth and spread of the fungus.

Hop and Strawberry Mildew (*Sphaerotheca humuli*).

This mildew or mould as it is often called belongs to the "true mildews," and therefore grows on the outside of its host. In wet seasons it has been known to do a considerable amount of damage. The first symptoms of the disease are small light-coloured spots on the leaves. In damp weather the white mycelium of the fungus soon appears and begins to spread rapidly. It also spreads from plant to plant by means of conidia very similar to those of *Erysiphe graminis*. After a time it becomes a greyish colour. In autumn the resting stage or spore-cases are formed; these are of a dark brown colour. The greatest amount of damage is done when the mildew spreads to the cones. The reddening of the cones by the spore-cases has given the name of "red mould" to this stage.

The spore-cases are very similar externally to those of *Erysiphe graminis*. Internally they are different as they contain only a single spore-sac or ascus, as may be ascertained by squashing one under a coverslip. This fungus also causes considerable damage to strawberries. This mildew also lives on several weeds, such as dandelions, meadow-sweet, and wild geraniums, but it has not yet been shown that hop plants can be infected from these weeds.

Remedial Measures.

Its position on the outside of the plant offers easy means of attacking this fungus. Sulphur in various forms has proved very useful. There are several fungicides used against this mildew:

(1) Flowers of sulphur,
(2) Liver of sulphur,
(3) Lime sulphur,
(4) Bordeaux mixture.

It is important to spray the plant early, i.e. as soon as or even before the mildew makes its appearance. Later spraying depends on the weather. All badly diseased plants should be burnt.

American gooseberry mildew, *Sphaerotheca morsuvae*, is a member of the same genus as the hop mildew. This disease can be controlled by two sprayings with Lime sulphur. For the details of this see Ministry of Agriculture's leaflet.

Rose mildew (*Sphaerotheca pannosa*) which causes so much trouble to rose growers is a suitable form on which to try experiments. Watch some rose trees as soon as the leaves are out for the appearance of mildew and when it appears treat them in the following manner.

Leave Set I untreated for comparison with Sets II and III. Dust the trees of Set II with a coating of flowers of sulphur (the leaves should be damp when this is done). Spray the trees of Set III with a solution containing 1 oz. of liver of sulphur to 3 gallons of water. Watch the trees carefully to see if the sulphur or liver of sulphur has any bad effect on the leaves or buds, and also to see if either of them prevents the mildew from spreading.

CHAPTER V

ERGOT AND CLOVER SICKNESS

Ergot (Claviceps purpurea).

This disease is caused by the fungus *Claviceps purpurea*. Its commonest hosts in this country are rye, rye grass, couch, cocksfoot, and timothy, and it is found occasionally on wheat and barley. Some of the grasses, such as the bromes, are not subject to it. The disease first shows itself at the time when the corn is ripening or the grasses are dying off. It is often abundant on the dried heads of roadside grasses. It appears as long blackish bodies sticking out from the ears (see Fig. 19). These are called Sclerotia or Ergots. They are partly covered by the chaff and take the place of the grain in the flower. The flowers of rye are much larger than those of the grasses, and the sclerotia produced in them are correspondingly larger. In the case of rye the sclerotia are often more than 1 inch in length and about $\frac{1}{8}$–$\frac{1}{4}$ inch in diameter. In the case of grasses they may be $\frac{1}{4}$–$\frac{1}{2}$ inch long, and about $\frac{1}{12}$ inch in diameter. The longer ones are usually curved but the short ones are often straight. They are hard and have a groove running along one side.

In appearance they are not unlike small thin date stones, but they are rather darker in colour. A cut across one of them shows that only the outer layer is

Fig. 19. Cocksfoot and rye grass attacked by *Claviceps purpurea*.

dark in colour, the interior being whitish, resembling very closely a cut grain of rye. Each sclerotium consists of hyphae matted tightly together and running in all directions, so that sections across it in any direction have the appearance of circular or oval cells, these being the cut ends of the hyphae.

Sclerotia sown just under the surface of damp earth or sand and kept in the open air begin to germinate in the spring. Small lumps at first form on them and eventually these take the form of short drum-sticks, the thickened portions being carried above the ground (see Fig. 20). On cutting sections these heads are found to contain a number of flask-shaped cavities in which are a number of club-shaped bodies known as asci, which in turn carry eight needle-shaped spores known as ascospores.

Fig. 20. A sclerotium of *Claviceps purpurea* which has germinated.

In the ordinary course of events some of the sclerotia fall to the ground and others are harvested. Germination of the sclerotia takes place in the spring in the ground. Some of them fall to the ground naturally and others are often sown with the seed. The ascospores are produced about the same time as the host plants flower. The spores are carried about by various agencies and some of them reach the flowers of the host plants. Here they germinate, putting out germ tubes which penetrate the flowers at their base. The mycelium formed from the germ tube grows through the base of the flower and then spreads over the surface covering the portion which normally

forms the grain with a network of hyphae. The free
ends of these become swollen and conidia are formed at
these swollen ends. These are set free and others are
formed at the ends from which they have fallen. The
hyphae also secrete a sticky sugary substance in which
the conidia become embedded. This is known as the
honey dew stage because of the small drops of sticky
substance which are formed. This sugary substance
is very attractive to insects. They carry it about with
them and so transfer the conidia to other flowers.

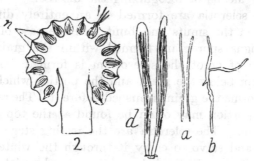

Fig. 21. *Claviceps purpurea.* (2) Section of fructification showing
(*n*) the cavities containing asci; *d*, asci; *a*, ascospores; *b*, germinating
ascospore. (Magnified.) (After Percival.)

A conidium which in this way reaches the flower of a
susceptible plant sends out a germ tube which penetrates
the base of the flower, a mycelium is formed similar
to that of the parent from which the conidium was
formed, and in its turn this very soon produces a crop
of conidia. A considerable number of flowers may
thus become infected from a single sclerotium. A large
number of spores are carried to flowers which they are
not capable of infecting and here they die. The fungus

forms a large number of conidia so as to increase the chances of reproducing itself.

After a time the mycelium at the base of the flower grows to form a compact mass. This is composed of hyphae woven together and running in all directions, the outer layer of which becomes darkened. This developing sclerotium pushes before it the portion of the flower which forms the grain and also the remains of the hyphae which bud off the conidia. It then increases in length and sticks out from the ear providing an easy means of recognizing the disease.

The sclerotia are formed in an entirely different way from the smuts and bunt. In the case of these the fungus spores are formed within the grain. In the case of ergot the sclerotium is formed from the mycelium below the grain and the portion which ordinarily forms the grain is pushed before it. The remains of this portion may often be found at the top of the sclerotium. The sclerotia are the resting stage of the fungus and serve to carry it through the winter.

This fungus is divided up into several varieties, the commonest of which attacks rye and several of our grasses.

The damage done to grasses is very little, but it may reduce the yield of rye considerably. This direct amount of damage is negligible compared with the harm it does when the sclerotia are eaten. They contain several poisonous substances. Cattle feeding on infested pastures during the winter often do badly and become very thin. Sores are often formed on the teats and mouth, the latter causing salivation. These symptoms of the poisoning effect have led to its being confused with the dreaded foot and mouth disease.

The poisoned animals often become lame. In very bad cases the extremities, such as the tail, hoofs, ears, and teats, may dry up and fall off. Another effect of feeding on ergot, which is very important from the farmer's point of view, is that it often causes abortion in cows fed on infected pastures.

In Russia and countries where rye bread is eaten this fungus brings about the above ailments in human beings and a number of lives have been lost through it.

If the sclerotia are kept for some time they gradually lose their poisonous properties.

Remedial Measures.

The sclerotia of this fungus are often present as impurities in seeds. They are sometimes mistaken for rats' droppings, but by breaking them it is easy to distinguish between the two as the sclerotia are whitish inside. All seeds should be carefully examined and any ergot present should be screened or picked out.

This disease is not noticed in its earlier stages, as it works hidden away under the chaff of the grasses or rye. When noticeable the sclerotia are already formed. In the case of infected pastures cattle should be taken off immediately and the grasses which have seeded should be cut, collected, and then burnt. After this the pasture must be regarded as an "ergoted" one and grasses should not be allowed to go past the flowering stage. If they are cut just as the flowers are forming the fungus is kept down considerably as the sclerotia are prevented from forming. At the same time those in the hedges must be cut at the flowering stage.

Meadow hay is perfectly safe for cattle as the ergot does not form for some time after the stage at which the grasses are cut. The grasses in the hedges and roadsides should be cut and burnt in late autumn.

Fig. 22. A clover plant attacked by *Sclerotinia trifoliorum*; white sclerotia are present at the top of the root. Later on these sclerotia become black.

Clover Sickness.

The dying off of clover, known as "clover sickness," is intimately associated with two organisms—one a fungus known as *Sclerotinia trifoliorum* (this disease is now called Clover Stem-Rot), the second a small eelworm known as *Tylenchus dipsaci* (see p. 166). The disease is usually noticed first on young clover in the autumn. Mild weather seems to be the most suitable

for the fungus. It has been found on red, crimson,
white, alsike clovers, and on beans, lucerne, sainfoin
and black medick. It causes much more damage to
red clover than to the others. The shoots of the
attacked plants droop, the leaves turn brown and usually
become matted together in a rotting mass. This state
of affairs spreads from plant to
plant in suitable weather. Small
black irregular bodies, something
like rats' droppings, are found on
the stems or roots or on the soil.
These are the sclerotia or resting
stage of the fungus. They consist
of a dense mass of hyphae (see Fig.
22). When kept in damp earth in
the open air they eventually[1] send
up small brownish saucer-like bodies
on the ends of stalks of varying
lengths (see Fig. 23). Near the
upper surface are situated a number
of asci, each of which contains eight

Fig. 23. *Sclerotinia trifoliorum*, a sclerotium
which has germinated
and produced a saucer-
like structure which
bears the asci. (After
Massee.) See also Fig.
55, p. 174.

oval ascospores. In nature these ascospores germinate
and produce a mycelium which enters the clover plant,
ramifies throughout the stem and leaves, and kills them.
After a time this mycelium forms sclerotia.

Remedial Measures.

There is no evidence that the disease is ever carried
by the seed. As the sclerotia are capable of living for
four or five years red clover should not be grown too

[1] Usually in October or November.

frequently on the same land. Sainfoin, Trefoil, or White Clover which suffer very little from this disease may replace red clover. Peas and tares may also be grown. Crops other than leguminous ones are not attacked.

CHAPTER VI

RUSTS

The group to which rusts belong, known as the Uredineae, is very widely distributed, in fact some of its members are present in all parts of the world where plants are cultivated. The term "rust" in its popular sense includes all diseases which cause a rusty appearance of a green plant. From a farmer's point of view they are extremely important, as they bring about a considerable loss in his cereal crop every year. They are characterized by giving rise to several different kinds of spores, some having as many as five different kinds. Another peculiarity of certain members of the group is that some of these spores may be borne on one host, and other spores of the same fungus on an entirely different host. There are a number of rusts which attack our cereals, most of them belonging to the genus *Puccinia*. The differences between them are chiefly concerned with the number of kinds of spores, their characteristics, and whether they are all or only in part produced on the cereal host. A description of two of the more important species will help to give some idea as to what rusts are.

Puccinia graminis. The popular name of this

fungus is black or orange rust of wheat, and it appears commonly in wheat fields in late spring or early summer. The leaves of the wheat first of all lose their characteristic green colour and become slightly yellowish. Dark orange patches are then formed on them, and are found between the veins of the leaf or on the stem. This is the first spore stage that the fungus produces on wheat and it is this which gives the popular name of rust.

Examination with a lens shows that these patches are composed of a number of minute bodies or spores clustered together, and that they are partly covered at the sides by the colourless skin or epidermis of the wheat leaf. They are known as uredospores or summer spores (see Fig. 24). On shaking the leaf some of them fall out as a powder. They are ovoid in shape, the coat or wall being of moderate thickness and covered with a number of very short blunt spikes. Midway between the apex and the point of attachment are some thin places in the wall. A section through one of the orange patches shows that each spore is borne on a short stalk (see Fig. 24). The mycelium which produces them does not extend throughout the whole plant, but is confined to small areas near the spores. It consists of a dense web of very fine hyphae growing in the spaces between the cells. Small tubes or haustoria grow into these cells for the purpose of obtaining food. The hyphae have occasional transverse walls. If the uredospores are placed in a hanging drop under suitable conditions for germination they send out tubes through the thin places in their walls (see Fig. 24).

One of these usually grows much longer than the others and branches. An uredospore falling on a blade

Fig. 24. *Puccinia graminis*. (1) Uredospore stage on wheat; (2) uredospore stage highly magnified; (3) uredospore stage very highly
magnified; (4) uredospore germinating; (5) teleutospore stage on
wheat; (6) teleutospores; (7) a teleutospore germinating; (8) aecidiospore stage on barberry leaf; (9) aecidium cup showing the chains
of aecidiospores; (10) two aecidium cups. (After Massee.)

of wheat germinates in the same way. When the germ
tube reaches a breathing pore it enters and so infects
the plant. The mycelium from an uredospore is said
to produce fresh uredospores in about 7–10 days.

These are continually produced throughout the
summer. In late summer we shall notice that the
appearance of the rust changes. The leaves look
darker and on closer examination this is seen to be
due to the production of dark brown or nearly black
spores together with the orange coloured ones. After
a time the latter disappear and the straw is now
marked with black lines, due to the long patches
which are placed one above the other in the soft tissue
between the veins. It is this stage that gives the name
of black rust to the fungus.

The black spores are not of the same kind as the
orange coloured ones and are not derived from them
by change of colour. They are known as *teleutospores*
or winter spores, and are produced from the same
mycelium which gave rise to the uredospores. After
successive crops of the latter the fungus suddenly starts
to form teleutospores, and so we get the change from the
orange to the black colour. A teleutospore is almost
twice as long as an uredospore and is divided into two
almost equal cells by a thick partition across it. Its
walls are very thick and smooth (see Fig. 24). They
take much longer than the uredospores to germinate
and on doing so each of the two cells sends out a tube
from its upper part. This becomes divided into three
or four cells by cross partitions; from each of these a
short tube grows out and becomes swollen at the end,
and eventually cut off as a small spherical body,
known as a secondary spore (see Fig. 24).

The fungus has now reached the stage when it leaves the wheat plant and seeks a new host. Before its life history was known it was held by farmers that barberry hedges caused the wheat near them to become rusted. This story aided de Bary in following out its further development. He found that it was possible to infect the barberry plant by means of the secondary spores formed from the teleutospores. He also found that the fungus produced on the barberry had a totally different appearance from the rust of wheat which produced it, but on infecting wheat with the spores from the barberry it became rusted. Two kinds of spores are produced on the barberry but only one of them takes part in the reproduction of the fungus. These are known as aecidiospores, and are borne in a chain-like manner. A series of these chains is enclosed in a kind of ball near the lower surface of the barberry leaf. After a time the ball bursts and a kind of cup is produced which can be seen with the naked eye (see Fig. 24). The aecidiospores are yellowish, spherical, and almost as large as the uredospores. On germination they send out a simple tube which is capable of infecting wheat. The aecidium cups appear on the barberry in the early spring. This is not a case of one fungus giving rise to another fungus as the aecidiospores cannot infect other barberry plants. The different spores are stages of the fungus *Puccinia graminis*.

The damage done to the wheat plant consists in reducing the yield, as some of the food which should have gone to swell the grain is taken by the fungus, and the amount of food manufactured is lessened by the destruction of some of the chlorophyll in the leaves and stem. In very bad attacks the grain may become shrivelled.

In cases of attacks by some of the fungi which live alternately on two hosts it is possible to reduce the amount of damage by the removal of one of the host plants. In the case of *Puccinia graminis* this plan does not succeed. In Australia there are no barberry bushes and no plant has been found on which the aecidium stage occurs and yet this fungus survives and does considerable damage. This is also true of certain parts of England. In this case it is supposed that the uredospores live until the next season's crop is growing and then infect it. It is possible that the mycoplasm hypothesis put forward by Eriksson in which he states "that the germ of the rust is present in the seed of the host plant" may have some foundation, but it is not usually accepted. This fungus has several different varieties which all appear the same under the microscope, but each will only infect certain host plants. That usually found on oats will not attack wheat or barley. That found on wheat, however, is occasionally found on oats and barley.

From our point of view the *Puccinia graminis* attacking our cereals may be divided into three different varieties: (*a*) usually attacks wheat and occasionally barley, oats and rye; (*b*) usually attacks barley, rye, couch grass, and other grasses, but does not attack wheat and oats; (*c*) usually attacks oats, tall oat grass, and cocksfoot, but not wheat, barley or rye.

There are other varieties which attack grasses but do not attack cereals.

Puccinia glumarum.

This species, known as yellow rust, is the commonest form of rust found in this country. It may be readily distinguished from *Puccinia graminis* by the uredospore stage. In this case the uredo patches or sori are orange yellow instead of dark orange coloured as in *Puccinia graminis* and most of the other rusts. The sori are smaller and are often united with their fellows, making them appear as lines on the leaf and stem; they are usually present in larger numbers. The uredospores themselves are more rounded, but otherwise they are very similar. The teleutospores are produced in smaller sori, and the deep gaping wounds of *Puccinia graminis* are never present. They are also much shorter and appear more like black dots than lines.

No aecidiospore stage has been found and no alternate host. The uredospores and teleutospores germinate in a similar manner to those of *Puccinia graminis*, and the uredospores probably carry on the disease from one season to another. The yellow rust of wheat cannot be distinguished from the yellow rust of barley under the microscope, but by experimenting it has been found that barley cannot be attacked by yellow rust of wheat, neither can wheat be attacked by yellow rust of barley. From our point of view this fungus may be divided up into a number of species each of which has its particular host or hosts.

There are numerous other rusts besides the two described, some of which possess all the spore stages and others only one or some of the stages; the differences are very similar to those of the two species described.

Remedial Measures.

All attempts to check rust by spraying and treatment of seed have failed entirely. The cereal plants
do not lend themselves to spraying as the nature of their
leaves is not suitable for holding the spray, but rather
the contrary. It is chiefly with yellow rust of wheat,
as being the commonest, that work is being done to
help the farmer.

The susceptibility of wheat varies very largely
among the different varieties. Some are practically
immune, but these are either bad yielders or of poor
quality. Biffen has found it possible to cross a heavy
yielding variety susceptible to yellow rust with one that
is immune and to pick out from the offspring a variety
with the cropping power of the susceptible variety and
the immunity of the other.

Wheat immune to yellow rust may be susceptible
to other rusts.

On crossing a wheat which is susceptible to yellow
rust with a variety which is immune from that rust,
the whole of the plants derived from the seed produced
are readily infected. On sowing the seed from these
rusted plants it is found that although the majority
of the plants produced are infected, about one-quarter
of them are free from it. Of these a certain proportion
possess the high yielding property of the original
rusted parent together with the absence of liability to
rust.

On growing these wheats the plants produced are
all practically immune from the yellow rust. In this
way a wheat known as Little Joss was produced by
crossing Square Head's Master with a wheat resistant

to yellow rust and known as Girka. Little Joss is resistant to rust and has a rather higher cropping power than the high yielding parent Square Head's Master. This increase of yield in the Little Joss over Square Head's Master is probably accounted for by the rust preventing the latter from yielding as high as it would do if it were not rusted. Little Joss is immune only to yellow rust.

It is to crossings of this kind that we look for assistance in helping to prevent the rust fungi from taking their yearly toll from our cereal crops.

Beet root Rust. The rust which attacks mangolds, beetroot, and sugar beet does not belong to the same genus as cereal rusts. It is known as *Uromyces betae.* It also occurs on wild beet.

The three stages are found on the same plant and it differs from *Puccinia* in that the teleutospores or winter spores are single-celled.

In spring the yellow aecidium cups may be found on the leaves. In summer the brown patches of uredospores are present and in autumn these change to a dark brown colour caused by the teleutospores. This disease is most prevalent in crops manured with too much nitrogen and not enough potash as is well shown on the mangold field at the Rothamsted Experimental Station. Diseased leaves should be collected and destroyed.

Pea Rust is caused by *Uromyces pisi.* The aecidiospore stage occurs on a wild spurge.

Bean Rust is caused by *Uromyces fabae.*

CHAPTER VII

SMUTS

On walking through a cornfield just before the grain is ripe here and there black ears, known as smuts or Black Heads, may be noticed, as they are easily seen by contrast with the colour around them. The blackening of these ears is caused by various fungi and it is from this black colour that they derive their name "smuts." They may be found on barley, oats, wheat, and also on a number of grasses. There are many kinds, most of which belong to the genus *Ustilago*. A host plant may have more than one species capable of attacking it, but each smut has its own particular host. Those attacking barley cannot infect the oat plant, neither can those found on oats attack barley.

In place of the grain which is present in a healthy plant, the blackened ears contain a dark brown powdery mass (this powder is very dark and causes the ears to look black). It consists of a large number of almost spherical bodies, which are the spores of the fungus. The development of these can be watched in a hanging drop of water. After a time a small germ tube is sent out. This does not grow very long but becomes divided up into four cells by three cross walls.

From the upper portion of each of these cells a process is given off, which eventually becomes divided from the cell producing it. These are secondary spores and in such places as manure heaps, where food supply is abundant, they increase in number by budding, but it is sufficient for our purpose to know that the plants are infected by means of these.

If grains of oats taken from a badly smutted field are sown, some of the resulting plants will be smutted and others quite clean. At the various stages of growth very little difference is noticeable between the plants which will produce healthy and those which will give smutted grain until well on towards harvest. The infected plants are said to be more robust and usually taller. When the grains are beginning to form some of the ears show a darker appearance, and as they ripen become blacker until a stage is reached when the ears seem to be making soot instead of grain (see Fig. 25).

In looking for a suitable means of dealing with the smuts it is very important to know how the plant is first attacked. Barley, oats, and wheat show differences in this respect.

Host.	Fungus.	Common name.	When infection takes place
A. Barley	*Ustilago hordei*	Covered smut of barley	
	" *avenae*	Loose smut of oats ...	At seedling period.
Oats	" *levis* or *kolleri*	Covered smut of oats ...	
B. Barley	*Ustilago nuda*	Loose smut of barley ...	At flowering period.
Wheat	, *tritici*	Loose smut of wheat ...	

These fungi are very similar in structure and at one time were considered to be the same. Experiments have shown that under no circumstances can either produce any of the others, thus indicating that they are all distinct forms.

Loose smut can easily be distinguished from covered smut as its spores come out of the grain and often leave the stem of the ears bare, whereas in covered smuts the spores remain in the grain on the ears.

In the case of the covered forms a large number
of the smutted ears are harvested with the sound ones.
On threshing the smutted grain is broken and the spores

Fig. 25.   Oat plants attacked by *Ustilago avenae*.

are spread to the sound grain. With loose smut of
oats the spores are blown about by the wind from the
time that grain is ripening until harvest. A number
of them settle on the chaff of the oat flower when it
opens, and on harvesting these get on the grain. In

the case of group *A* the spores are sown with the grain. Conditions which favour the germination of the oats and barley apparently favour the germination of the spores. The secondary spores formed from these produce germ tubes which make their way into the shoots of the seedling plants. It has been shown that the fungus can only obtain an entry into the plant at a very young stage. When the leaves have pushed their way through the leaf sheaths the shoots become hardier and spores cannot enter.

On entering the plant the germ tube, or as it is now called the mycelium, makes its way to the growing point of the shoot. It takes its food from the cells of the plant but does not kill them as in the case of the potato disease; consequently there are no external signs that the plant is being attacked. As the oat grows in the spring the mycelium grows with it and is to be found in the region of the growing point, but not in the older parts of the plant. By maintaining its position at the growing point it is able to pass out into the new shoots as these are all formed there. In this way all the flowering shoots of the attacked plant are usually infected and practically all the grains of an infected ear are smutted.

When the plant is supplying the grain liberally with food stuff and so causing it to swell the mycelium grows rapidly and a dense network is formed within the grain. The spores are formed inside the tubes of the mycelium and set free as a powder.

The loose smut of wheat appears at the time of flowering. The coverings which enclose it burst about this time and the black spores are blown about by the wind or washed to the ground by rain. Soon nothing

remains but the main stalk or rachis of the ear which
usually has a certain amount of smut adhering to it.
Some of the spores reach the stigmas of wheat flowers.
Here a germ tube is produced which grows through the
style much in the same way as the tube from a pollen
grain. Eventually the tube reaches the developing
seed. The grain so infected ripens normally and can
in no way be distinguished from a non-infected grain.
When these grains containing the smut mycelium are
sown the ears of the plants which they produce become
smutted and spores are blown about to infect fresh
plants. The mycelium maintains its position at the
growing point as described above. The loose smut
of barley has a similar life history. The weather
plays a big part in the prevalence of the loose smuts.

If at the time the spores are forming there is a lot
of rain the majority of them are washed to the ground,
with the result that fewer grains become infected and
less smut is produced in the following season.

Remedial Measures.

In the life history of the smuts we have seen that
they may be divided into two classes: (*a*) those which
infect at the seedling stage; (*b*) those which infect at
the flowering stage. The treatment will therefore be
different in the case of these two groups.

In both the disease is spread from smutted ears,
and consequently in choosing seed corn we shall do
well to obtain it, if possible, from fields which contain
no smut. This does not ensure our crop from attack
as the smut may have blown from neighbouring
fields, but seed from a clean field usually produces
considerably less smutted ears than that from a

smutted field unless measures are taken to kill the spores.

In the case of group A, which attacks the host at the seedling stage, the spores are sown with the grain. Lines of procedure consist in employing some method which will kill the spore and not injure the grain. Numerous experiments have been carried out with this object in view, and from them the following methods have been shown to give satisfactory results.

In Method 1 blue stone (copper sulphate) is used to kill the spores. A clean wooden or concrete floor is best for this treatment. A sack (i.e. 4 bushels) of grain is emptied on this floor. The copper sulphate solution is made by dissolving $1\frac{1}{2}$ lbs.[1] of copper sulphate in $1\frac{1}{2}$ gallons of soft water. The grain is then thoroughly moistened by turning it over and watering it with this solution. It should be turned several times and allowed to stand for about 10 hours. It is then spread in a thin layer until dry, when it is fit for sowing.

Another way of using copper sulphate is to dissolve $2\frac{1}{2}$ lbs. of it in 10 gallons of water and pour the grain into this, leaving it there for about 14 hours. The copper sulphate is then run off, and the grain is spread to dry. This method kills the smut spores, but it also hinders the germination of the grain, especially of oats and barley, and should not be used for them. The loss is said to be lessened by watering with milk of lime, or dusting with lime after treatment.

Any seed left over after sowing should not be used for feeding purposes as the copper sulphate is harmful.

In Method 2 commercial formaldehyde is used to kill the spores. It contains nearly 40 per cent.

[1] One quarter of this strength has been found effective and less injurious to germination.

formaldehyde. One pint of it is mixed with 30 gallons of water. The grain is then poured into a basket which stands in the solution and continually moved about for 10 minutes. It is then spread to dry as rapidly as possible. The same strength of formaldehyde is often used to sprinkle[1] on the grain in the same way as with copper sulphate. This is the most satisfactory dressing yet tried. It does not injure the germination so much as treatment with copper sulphate. Treated grain if properly dried may be fed to fowls.

In Method 3 warm water is used to kill the smut spores. Two tubs, several buckets, and a supply of hot and cold water are required. It is essential that an accurate thermometer be used. Hot and cold water are put into the two tubs. The temperature of the water in Tub 1 should be about 125° F., and in Tub 2 132° F. (not below 130° F. and not above 135° F.). The grain is placed in a fine wicker or gauze basket or in a sack and moved about in Tub 1 for 2 or 3 minutes. It is transferred to Tub 2 and kept there for 5–8 minutes, and then shot on a raised piece of canvas and sown as quickly as possible or dried rapidly.

Hot water must be added to the tubs from time to time to keep the temperature correct. Care must be taken that it is not poured directly on the seed.

A temperature of 132° F. is sufficient to kill the smut spores, but it does not injure the grain.

This method is very efficient and cheap, but is not suitable for farm use owing to the care required in keeping the temperature correct. If the temperature is too high the grain will not germinate, and if too low the smut spores are not killed.

After treatment with any of these methods the

---

[1] After sprinkling the heap should be covered with sacks (dipped in diluted formaldehyde) for about four hours and then spread to dry overnight.

corn must be put into clean sacks or bins. Sacks may be cleaned by boiling or the bins by washing with a solution of formaldehyde. The grain must not be allowed to freeze if it is at all damp. If sown immediately after treatment a larger bulk will be required, as it will be somewhat swollen.

Get 100 oat grains which have been grown in a field almost free from smut; and also some smutted ears of oats (either loose or covered smut). Shake up 80 of the grains with the smutted ears so that the oats become covered with the smut spores. Treat the infested grain as follows: (1) twenty with copper sulphate as described in Method 1; (2) twenty with formaldehyde as described in Method 2; (3) twenty with hot water as described in Method 3.

Now sow the 100 grains in pots, making five sets of twenty plants. Set (1) uninfested and untreated; (2) infested and untreated; (3) infested and treated by the copper sulphate method; (4) infested and treated by the formaldehyde method; (5) infested and treated by the hot water method.

The plants should be examined in all stages and any differences, such as time of appearance above ground and size of plants, should be noted.

We shall expect to find that Sets 1, 3, 4 and 5 are free from smut and that all the ears of the plants in Set 2 are covered with smut.

This experiment shows that oats can be infected with smut from spores on the grain but that these spores can be killed with little harm to the grain by either of the three methods used.

Now take some wheat grains and cover them with loose smut of wheat and sow them as before. These grains grow perfectly free from smut as the loose smut of wheat is incapable of infection at the seedling stage.

The public threshing machine is a great friend of those smuts which infect at the seedling stage, as it serves to spread them from one farm to another. When smutted corn is threshed the next lot that goes into the machine will be infected unless the thresher is cleaned. If used it should be thoroughly cleaned beforehand. The drill should also be thoroughly cleaned.

In the case of the group *B* the smut fungus is already present in the harvested grain, having got in at the flowering period. Dressing the grain as for group *A* is not a prevention as the steeps cannot reach the fungus. A modified hot water method has been found which kills the fungus and at the same time does little damage to the grain. This method consists of soaking the seed for about 6 hours in cold water and then putting it for 10 minutes in water at 129° F. for wheat and 13 minutes at 126° F. for barley. The soaking in cold water apparently affects the fungus in such a way that the hot water kills it.

These two smuts are present in most of our wheat and barley fields to a much greater extent than the covered forms, as the latter are usually prevented by one of the three methods given above.

Bunt or Stinking Smut.

Wheat is the only cereal which suffers from this disease. The diseased plants are very similar to healthy ones until the approach of harvest. On passing through a wheat field, containing plants which are bunted, by observing closely we shall be able to note that in some of the ears the wheat grains are visible. These bunted grains are swollen and cause the glumes or chaff to gape. Their appearance varies with the different varieties of wheat. In most it is short, swollen, and almost spherical. In some, however, instead of shortening it becomes longer. The groove is never so pronounced as in healthy grains and in some cases is very shallow. The bunted ears are usually darker as the grain is greyish or greyish black.

By rubbing an infected grain between our fingers, we shall be provided with the means of recognizing this disease on future occasions. It gives off a very disagreeable odour—somewhat resembling stinking fish—which clings to the fingers for some time.

It does not attack the chaff but only the interior of the grain. The coat of the grain is not broken, and inside are many dark-brown spores. There are two species of bunt, *Tilletia caries* (see Fig. 26), which has rough spores, and *Tilletia levis*, which has smooth spores.

In many cases all the spores are rough, indicating the presence of *Tilletia caries*, but in some cases the smooth spores of *Tilletia levis* may be mixed with them and in a few cases all the spores may be smooth. As these two species behave in a similar manner, we shall confine ourselves to the former. The spores are almost spherical and their surface has a rough, warty appearance. They are very small, being about $\frac{1}{1500}$ of an inch in diameter. In the case of this fungus the diseased grains are harvested with the good ones and on threshing some of them are crushed with the result that the spores get on the sound grain. When this wheat is sown the spores germinate, and eventually the young wheat plant is infected. The brown spores yield a crop of secondary spores, and these

in their turn give rise to tertiary spores from which an
infection tube is produced. It is sufficient for our
purpose to know that the spores adhering to the wheat
grain germinate and eventually infect the young wheat
plant. They are said to retain their capacity for
germination for seven or eight years. When the young
plant is infected the mycelium grows as in the case of

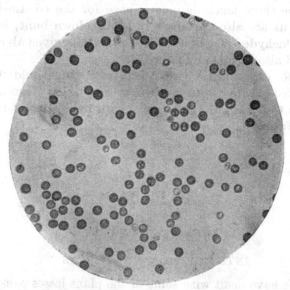

Fig. 26. Spores of *Tilletia caries* as seen under a microscope.

Ustilago, without killing the cells of the host, and the
spores are formed within the mycelium. Usually all the
grains of the attacked plants are filled with spores. The
damage done by this pest is not only the loss of yield due
to the damaged grain for this is only small in proportion
to the total, but on grinding wheat which is infected the
flour has a dirty appearance and the peculiar odour of

the fungus is very noticeable. It lowers the price of the flour even if present only in small quantities, and is sometimes sufficient to render it unsaleable. There is no evidence of infected grain causing any damage when fed to stock.

Remedial Measures.

The three methods given above for the treatment of smuts are also effective in keeping down bunt, but formaldehyde is the best. The precautions given above should also be closely followed.

The experiment with smut of oats should be repeated, using wheat grains for sowing and bunted grains to infect the clean grains with the spores of the bunt fungus.

PART II

CHAPTER VIII

INTRODUCTION TO INSECTS

We have dealt with some of the plant losses caused by fungi and will now turn our attention to those caused by members of the animal kingdom. By far the greater number of these losses are due to the ravages of insects.

Popularly most small animals such as flies, beetles, butterflies and spiders are considered as insects. Scientifically animals are classified into groups according to their structural characteristics. The following

characteristics distinguish insects from other animals:
the bodies of the adults are divided into three parts,
the front known as the *Head*, the middle as the *Thorax*,
and the hind portion as the *Abdomen*.

The head bears: (1) The eyes, which may be (*a*) simple
or (*b*) compound eyes, composed of what look like a
large number of hexagonal cells. (2) The *antennae* or
feelers, which vary considerably in different insects,
and may be simple pointed rods or may possess knob-
like or feathery endings. (3) The mouth, which is an
opening below or in front of the head around which are
grouped the jaws. These are fashioned according to
the functions which they are called upon to perform.
In biting insects they work horizontally and not up
and down as in the case of higher animals.

The thorax or middle portion of the body is divided
up into three rings or segments, each bearing a pair of
jointed legs. Both the second and third segments
usually bear a pair of wings, a marked feature of the
group. In flies however only the first pair are present
on the middle segment, the third bearing the remains
of the second pair which are used as balancing organs
(see Fig. 45). In some species, e.g. certain ants and
springtails, no wings are present, but other characters
resemble those of other members of the group. It is
therefore correct to classify them as insects.

The abdomen, usually the largest portion of the
body, is composed of a number of segments which
closely resemble one another. They have no appendages except in certain cases at the hind end of the body.

Unlike the higher animals insects have no internal
skeletons to support and protect their bodies, but the
outside is covered with a horny substance known as

chitin, which serves the same purpose. Insects do not breathe in the same way as the higher animals, that is to say they have no lungs into which they draw air through a special opening. They have, running throughout their bodies, a number of air-tubes known as *tracheae*, which give off numerous branches serving the same functions as the lungs, their exterior openings, at the sides of the segments, being known as *spiracles*.

Differences in such parts as wings, mouth parts, antennae, legs, and eyes, etc., help us to distinguish one insect from another.

The above characters also enable us to decide whether the animals under observation are to be classed as insects. Examination of a butterfly shows that it possesses all the above characteristics and therefore must be classified as an insect. If the animal we are examining is a spider we find that its body is not divided into three parts, that the adult form has four pairs of legs and that it possesses neither wings nor feelers. These and other characteristics have placed spiders in a different group.

An insect differs very considerably from most of the higher animals in its structure and appearance at the time of its birth as compared with its adult stage. When a calf is born or a young chick hatched each is very similar in appearance to its adult form, only very much smaller. In the case of most insects this is not so. On hatching from its egg it has few of the adult characteristics. It passes through several changes during its life, its appearance at one stage being very different from that of the other stages. A description of the life of a typical insect will help us to understand these changes.

Most of us know the common large white butterfly
which flies about gardens and fields during the summer.
It is called the Large White Cabbage Butterfly, its
scientific name being *Pieris brassicae* (see Fig. 27).
If we watch this butterfly closely we shall see that it
sometimes settles on a cabbage or a turnip or some
allied plant and remains for a considerable time underneath one of the leaves. If we wait at some distance
until the butterfly flies away of its own accord and then
examine the leaf on which the butterfly has been, a
number of very small yellowish bodies arranged in a
cluster may be found. They are almost skittle-shaped
and marked by longitudinal ridges (see Fig. 27 *a*).

In about ten days these disappear and are replaced
by a number of small green worm-like creatures known
as caterpillars. The small yellow bodies are the eggs
of the butterfly, and were laid by the female. The
small green caterpillars hatch from the eggs, eat the
empty egg shells, and then begin to feed on the leaf,
leaving marks where they have been feeding. They
grow rapidly, but after a time growth ceases as
the hard chitinous skin only allows of a certain
expansion. At this stage the caterpillar fixes itself
on a small silken web and a new skin is formed under
the old one, the latter bursting and allowing the
caterpillar to crawl out. It again grows rapidly and
the new skin hardens. This process of casting the
skin is repeated four or five times and allows the
caterpillar to grow to a considerable size. When fully
grown it measures about 2 inches or more in length.
It is bluish green above and yellow below, having a
long yellow line along its back, and also one along
each side. These are broken by a number of black

Fig. 27 *a*.   Eggs of *Pieris brassicae*.

Fig. 27 *b*.   Caterpillar of *Pieris brassicae*.

Fig. 27 *c*.   Pupa of *Pieris brassicae* with butterfly emerging.

Fig. 27 *d*.   *Pieris brassicae* (female).

spots. The whole surface is covered with fine hairs. The body consists of a distinct head and thirteen segments. The head is of a dirty brown colour with black spots. Each of the three following segments bears a pair of jointed legs. The third, fourth, fifth and sixth, and also the last of the abdominal segments bear a pair of false legs each. These are simple protrusions from the lower surface of the body (see Fig. 27 *b*). The stage after the egg—in this case the caterpillar— is known as the larval stage and the young insects— caterpillars—are known as larvae.

When fully grown the caterpillars usually leave the cabbage and wander off to sheltered places where they change into the next stage. Here the caterpillar forms a silken web-like platform, fixes itself by its tail head downwards, or horizontally, also placing a silken thread round the middle of its body as additional support. Its skin is then cast, and eventually a hard body is formed incapable of movement except of its tail. This is the *Pupa* or *Chrysalis* stage, and is very different from the caterpillar (see Fig. 27 *c*). It is about an inch long, of a pale green colour, spotted with black, and angular. The divisions into head, thorax and abdomen can be roughly determined. After about a fortnight the adult butterfly comes out. This stage is known as the *Imago* stage. The adult—as is also the case with other insects—does not grow, a small fly never becomes a large one.

The front wings of the adult butterflies are white on their upper surface with a black crescent-shaped patch at the tip of each. The females may be distinguished from the males by the presence of two black spots in the middle and a small black patch at the hind edge of these wings. The under surface is white with a yellow tip and in the case of the female it has two black spots as on the upper surface (see Fig. 27 *d*). The hind wing is white on the upper surface with a small black patch on the front edge. Underneath it is yellowish. The colours of butterflies and moths are due to the presence of numerous scales.

After pairing the female proceeds to lay her eggs on a plant on which the caterpillars can feed. This butterfly has at least two broods every summer, that is to say the butterflies formed from the eggs laid in early summer give rise to caterpillars in the same summer. In favourable seasons three broods are formed. While in the caterpillar stage this insect is liable to attack from various enemies. Birds and wasps kill a considerable number. I have seen a single wasp kill five in half-an-hour. They are also attacked by a small ichneumon wasp which by means of a piercing organ lays its eggs in the body of the caterpillar. The young larvae hatching from these eggs feed on the inside of the caterpillar, thus causing its death, and eventually come out and form little yellow bodies or pupae (see Fig. 28). These bodies, which are falsely known as *caterpillars' eggs*, should not be destroyed for they develop into wasps which lay eggs in other caterpillars and prevent them from forming butterflies.

In gardens the caterpillars can be prevented from eating cabbages by searching for and destroying the eggs in July and August. As the eggs are laid in groups this is not very difficult. Any caterpillars which hatch out may be destroyed at the same time.

Most insects are similar to the cabbage butterfly in that they are derived from eggs. The old idea that insects may be derived from putrefying matter may easily be shown to be wrong by keeping such material as putrefying meat so that insects cannot get near it. If however the meat is kept so that insects can reach it,

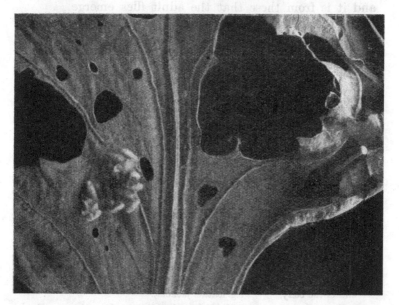

Fig. 28. A cabbage leaf bearing the yellow cocoons of an ichneumon wasp. (These cocoons are falsely called caterpillars' eggs.)

they will lay their eggs there, and from these maggots will be formed, and eventually flies.

The passage through these changes from the time they are laid as eggs until they become winged forms is known as the *transformation* or *metamorphosis* of insects. Three different stages of metamorphosis are found in all butterflies and moths and also in a large number of other insects.

In flies the stage after the egg is usually known as a Maggot. Those on fly-blown meat will serve as an example. These eventually change into brown packet-like bodies known as puparia, pupa-cases or cocoons, and it is from these that the adult flies emerge.

In their first stage beetles are popularly known as Grubs. The wireworm is one of the best known. It changes into a resting pupa stage from which the adult beetle is formed.

We often find insects in one stage only at one time so that it is important that we should be able to recognize them in all stages.

The following key gives a rough means of distinguishing the three large groups:

	Butterflies and moths	Flies.	Beetles.
Larva	A caterpillar possessing a distinct head, 3 pairs of jointed legs, 1–4 pairs of false legs on abdomen, 1 pair of false legs at the hind end of the body.	A maggot. No legs. No distinct head. Anterior end narrow, possessing a pair of dark jaws (some fly larvae have distinct heads, e.g. leather jackets).	A grub possessing a distinct head, 3 pairs of jointed legs or none. No false legs.
Pupa	The appendages and wings are encased. The tail is ringed and capable of movement.	Entirely enclosed in a case or cocoon (in the case of the leather jacket it has no case).	Appendages and wings free.
Adult	Two pairs of wings covered with scales are present.	One pair of wings is present. The position of the second pair is marked by "balancers."	Two pairs of wings are present: those of the outer pair are horny and do not overlap.

N.B. The larvae of certain sawflies closely resemble caterpillars but they possess more than four pairs of false legs on the abdomen.

In some insects such as aphides (green fly) and cockroaches, the young on hatching resemble the adult forms very closely—the difference being chiefly one of size.

The first consideration in dealing with pests attacking crops is to find out what insect is causing us trouble. Our knowledge of the different stages should help us considerably in at least finding out whether it is being caused by a butterfly, a fly, or a beetle.

The time taken by an insect to go through its different changes varies according to the species. In the case of wireworms and chafer beetles several years are required from the laying of an egg until the time when the insect formed from this egg commences egg laying. In some cases, such as the garden chafer, the insects have one brood every year, in others, such as the frit fly, several broods in the year, whilst green flies may have new broods every week.

The length of life of any one stage varies according to the species as does also the stage in which the damage is done. Often the larval stage only is responsible for this: in other cases it is done both in the larval and the adult stage, and occasionally in the adult stage only.

It is not necessary for the farmer to know what kind of insect is attacking his crop provided that he has sufficient knowledge to keep the pest under control. In order to deal successfully with an insect pest it is usually necessary for some one to study carefully the characteristic features of that insect throughout the whole of its life and very often throughout the life of some of its descendants. Such knowledge often enables us to attack the insect at a time when it is most vulnerable; otherwise treatment is haphazard and may mean attacking the insect at a time when it is most resistant, or sometimes attempting measures which prove to be useless.

The object of studying the life history of an injurious insect with a view of preventing damage is to find the weak periods in its life. We want to know the time and whereabouts of the insect when it is most easily kept in check. Knowing this we can find only by experiment the best means of controlling it. A knowledge of the cultivated plants and weeds on which the insect can feed is useful as it enables us to starve the insect or drive it away by keeping these plants from an attacked field for a certain length of time.

The life history tells us where the various stages in the insect's life are passed, how long it lives in these stages, and when the change from one to another takes place. From this we can often decide when and how to attack the insect in order to keep it under control.

In the case of the winter moth whose caterpillars eat the leaves of fruit trees the females have no wings, and when they want to lay their eggs on the young shoots they crawl up the trees. In England this takes place from about the second week in October until the middle of January. It was argued that if these females could be prevented from crawling up the trees the caterpillars would not be able to eat the leaves. An experiment was tried of putting a band of a sticky substance (as used on fly papers) round the tree trunks and keeping them sticky during the time that the female moths are active. It was found that when they reach the bands they stick and are unable to go farther.

They often lay their eggs on the bands or sometimes below them, but even if the caterpillars hatch they cannot travel the long distance to the leaves where they usually feed. From this experiment it was decided that the damage done by these caterpillars could be to a large degree prevented, and this method of grease-banding trees is now largely used in commercial fruit-growing. In seeking remedies for controlling insect diseases we must always bear in mind the question of expense. The farmer grows his crops for profit, and if it costs more to control the insect than the value of the damage done such measures are of no use to him.

One very annoying factor in dealing with insects is that they can sometimes travel considerable distances, so that a farmer who keeps them down on his own farm is often troubled by others coming from his neighbour's.

Insects may be roughly divided into two classes: (1) herbivorous, which feed on plants and cause diseases; (2) carnivorous, which feed on other insects and allied animals.

The members of the second group are extremely useful as they keep down the numbers of those which attack plants, e.g. large numbers of green fly are devoured by the larvae of lady-birds.

In nature both kinds live side by side in such a way that a sort of balance is set up so that no single species increases rapidly in numbers. In planting large areas with one particular crop man has upset this balance by providing the insects which live on this crop with an ideal feeding ground. Many herbivorous species are capable of living only on certain families or orders of plants. In nature this means that they often have to travel long distances in search of food, which in many cases they may be unable to find. In fields where we grow plants suitable for certain insects these can get food without searching for it and are thus enabled to increase rapidly. In this connection it must be remembered that insectivorous birds are useful in checking the increase.

Man has further upset the balance of nature by introducing foreign insects capable of living on our plants, e.g. the American woolly aphis. These are often capable of enormous damage as the insects, birds and animals which prey on them in their native country may not be present here. The introduction of the enemies of imported pests has in some cases proved extremely useful. Our present system of marketing also helps the spread of diseases from one part of the country to another.

Carnivorous species are capable only of keeping the herbivorous form within certain limits, and as man has favoured the increase of injurious insects he must, if he wishes to prevent them doing a considerable amount of damage, employ some means to reduce their numbers. A single insect is not capable of doing much damage, but it reproduces itself at an enormous rate and gives rise to numbers which collectively can often render crops almost useless. This rate of reproduction is necessary in nature because of the numerous factors which lead to the death of so large a percentage of the total number of eggs laid.

In endeavouring to keep injurious insects under control it is better to aim at prevention rather than at cure. Clean farming prevents a large number of insects from living through the winter as the weeds on which they would feed are destroyed. Dirty fields full of weeds and rubbish are ideal wintering places as they provide both food and shelter. The destruction of grasses and other weeds in headlands, hedges and ditches destroys many, and also gets rid of a favourite wintering place for others. The time of carrying out the various tillage operations affects the number of insects which are destroyed by such operations. Suitable rotations are very beneficial in keeping some pests in check.

The date of sowing often decides whether the crop will be good or poor. Thus the frit fly maggot shows a preference for young oat plants; early sown crops may therefore almost escape the attack, whereas late sown crops may be rendered almost useless. It is often possible to sow at such times that the insect is capable of doing the least amount of damage. Artificial manures are also useful in hastening the plant through the stages when it is most readily attacked.

Various substances known as insecticides have proved useful in killing certain forms which attack the shoots of plants. The kind to use depends on the way in which the insect feeds. Some, such as caterpillars, have biting mouths and feed by eating the leaves and shoots. The insecticide used in these cases is a poison, such as lead arsenate. This poison is sprayed on the plant so that on eating the insects are killed. Others, such as green flies, have sucking mouths and live by penetrating the tissues and sucking the juices. If a poison is applied on the surface of the plant it has no effect on them as they do not eat the surface layers. In this case an insecticide is used which kills the insects directly, i.e. by actual contact with it. A mixture consisting of soft soap and substances such as nicotine or paraffin is often used in such cases.

CHAPTER IX

BUTTERFLIES AND MOTHS

Surface Caterpillars.

There are a number of moths whose caterpillars damage many crops by feeding on them below the surface of the soil or at the soil level. This habit gives them the name of Surface Caterpillars. There are several different kinds, but the damage done by them is very similar and the means of keeping them under control are the same, so for our purpose we may group them together.

The chief offenders are:

Agrotis segetum ... ... The turnip or dart moth.
Agrotis exclamationis ... The heart and dart moth.
Triphœna pronuba ... The yellow underwing moth.

These do considerable damage to mangolds, turnips, carrots, parsnips, cabbages, potatoes, wheat and grasses. They attack most cultivated plants, but are partial to certain species.

They live as caterpillars during the greater part of the year. In the early summer they destroy young mangolds and turnips by cutting through the roots just below the surface of the soil before and after they have

been thinned. It is very noticeable after thinning as
gaps are made in the rows. The caterpillars may be
found by hunting in the soil near freshly damaged plants.
Later on in the season they attack the swollen roots of
turnips and mangolds by scooping out large holes. In
a similar manner potato tubers are attacked. The

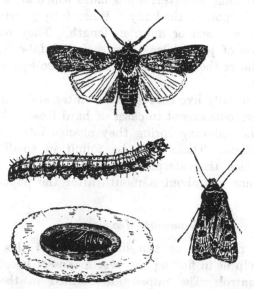

Fig. 29. Caterpillar pupa and adults of a surface caterpillar moth.

nibbled roots and potatoes often rot. Corn crops often
suffer from attacks on their roots in mild winters. The
grubs remain hidden under the surface or under leaves
during the day and come out to feed at night. The
moths giving rise to these usually fly at night and are
attracted by lamps or fires. Their hind wings are
lighter in colour than the front ones. The female has

7—2

a simple feeler, whereas that of the male is toothed. They are usually to be found in June, July and August. The eggs are laid on cultivated plants or weeds near the surface of the ground. The caterpillars are of a dull colour, either brown or greyish, and consequently are hard to find. They are cylindrical and apparently smooth, having, however, a few hairs which arise from small black spots on the body. When fully grown they are about an inch or more in length. They possess three pairs of jointed legs, four pairs of false legs in the middle of the body and a pair of false legs at the hind end.

They usually live through the winter and continue to feed on roots except in cases of hard frost. At the end of the following spring they change into smooth brown pupae. These may be found in small oval spaces which the caterpillars make in the soil. The moths come out about a month after the pupae are formed.

Remedial Measures.

Our knowledge of the life history of this pest should help us in finding the best means of keeping it under control. The pupae are formed in the soil when a crop is growing and consequently are not readily attacked except by the usual practice of hoeing. As the moths themselves are attracted by light it was thought that use might be made of trap lanterns. This however did not prove a success and is rather expensive as the period of hatching is long. We must then turn our attention to the destruction of the cater-pillars themselves.

They are difficult foes to combat owing to their

habit of living underground. In the case of drilled
crops such as turnips and mangolds, frequent hoeing
is found to be useful as some of the caterpillars are
destroyed and others exposed to birds. Starlings,
plovers, rooks, and gulls will eat them in large numbers.
Pigs will also eat them.

With turnips and mangolds sown on the flat
harrowing has sometimes proved effective.

In market gardens handpicking has been found to
be remunerative. The grubs are very difficult to find,
except after dusk. Each worker should carry a lantern
and a stout stick pointed at one end to search for caterpillars near the plants.

In the case of an attack on potatoes earthing up
as early as possible has proved effective. Weeds should
be kept down as eggs are often laid on them and also
because the young caterpillars feed on them until the
crop is ready.

Poisoned baits put along the rows have proved
advantageous. The bait consists of some green food
stuff, such as clover or lucerne, dipped into or sprayed
with a solution of Paris green (made by mixing 1 oz.
of Paris green with 3 gallons of water), or lead arsenate.
When the caterpillars eat this they are poisoned.

In fields of cabbages, mangolds, or turnips, where an
attack is feared, a small seed bed of cabbages will be
found useful as the blank spaces in the rows caused by
the caterpillars can be quickly filled by the cabbage
plants.

Toads will destroy enormous numbers and should
be preserved.

Other Caterpillars.

There are numerous other caterpillars (i.e. the larvae of butterflies or moths) which cause damage to our crops and especially to fruit trees. These may be distinguished from other larvae by the key on p. 92. Some of them are kept down by grease-banding the trees which prevents the wingless females from reaching the shoots. In other cases the trees are sprayed with a poisonous substance such as lead arsenate, so that when the caterpillars feed, they are poisoned. In addition to the caterpillars of the large cabbage butterfly we may find on cabbages those of the small cabbage butterfly eating holes in the leaves, and also the green or brown caterpillars of the cabbage moth (*Mamestra brassicae*) which eat into the hearts of the cabbages, leaving a lot of excrement behind them. As the small cabbage butterfly lays its eggs singly it is difficult to control.

The cabbage moth lays its eggs in groups and these should be looked for and destroyed in July and early August.

The caterpillars of the small swift moth (*Hepialus lupulinus*) are fairly common in gardens. They feed in much the same way as surface caterpillars on such plants as lettuce, potatoes, celery and bulbs. They are whitish in colour with brown heads. Scooped out potatoes will serve as traps.

CHAPTER X

BEETLES

Wireworms.

The term "wireworm" is used as a popular way out of the difficulty of giving a name to any small worm-like animal, and it often includes such different forms as leather jackets, millipedes, and centipedes. True wireworms are the larvae of beetles known as click beetles or spring-jacks. They attack all kinds of crops with the exception of mustard and are usually most abundant in permanent pasture, temporary leys, or any covered ground which is not disturbed. They do much damage to roots, corn crops, and peas, especially to crops grown on newly broken pasture land. They feed chiefly on or just above the roots of plants. It is not the amount of food eaten which makes them so destructive, but their habit of biting off the plant above the roots.

On pulling up a cereal that is being attacked, the wireworm may be found feeding on the stem a short distance from its base. The damage is done only by the wireworm or larval stage, but as it lives for three to five years the amount of harm done by a single individual is very considerable.

The eggs are laid in the soil at depths of $\frac{1}{4}''$ to $2''$ in June and July. The young wireworms which hatch out from these are very small and whitish. They begin to feed immediately after hatching. Full grown wireworms vary in size according to the species of beetle from which they are derived. Some may reach a length of 1 inch and others only ½ inch. The body is cylindrical and divided up into segments. In the course of their growth the skin is shed several times in the same way as that of the caterpillar of the cabbage butterfly. Their colour varies somewhat: after a moult they are usually pale yellow, but later on they become a rich golden yellow. The surface always has a shining appearance, and the skin is very tough.

The head carries a pair of powerful jaws which enable it to feed on all kinds of roots. Each of the three segments behind the head possesses a pair of jointed legs which readily distinguishes it from other so-called wireworms (see Fig. 30). The last segment of the body also has a downward projection which is used in walking. When fully fed the wireworm makes usually in the top 4″ an egg-shaped space in which it changes into a pupa. This change usually takes place in August and September and from the pupa the beetle hatches out in about 3 weeks. The beetles may often be found on grass land or clover leys during the summer. There are several species all of a dull brown or greyish colour, rather long and narrow, having short legs and serrated feelers (see Fig. 30 *C*). When placed on their backs they spring into the air with a clicking noise and fall on their feet. They are enabled to do this by means of a projection on the segment carrying the first pair of legs which fits into a depression on that bearing the second pair. This structure enables us to recognize the beetles fairly easily. They also make a clicking noise if held between the finger and thumb. The beetles are not

often found as they usually hide during the day time.
The wireworms feed throughout the whole of their
lives on the roots of either cultivated or wild plants,
except in frosty weather, when they go deep down into
the soil. After a thaw they come up again and feed.

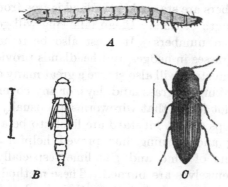

Fig. 30. Click beetle or wireworm beetle. *A*, wireworm; *B*, pupa;
C, click beetle. (About twice natural size.)

Remedial Measures.

The first remedy which suggests itself is if possible
that of preventing the eggs being laid. This has proved
of little use to the farmer as it cannot be successfully
carried out on grass land and clover leys. In gardens a
number of beetles may be trapped by means of small
heaps of clover covered with boards or tiles. The
beetles which shelter in these traps may be collected
at intervals and destroyed. This should be done in
April and May before the eggs are laid as the beetle
itself does no damage and no advantage is gained if
they are trapped after laying their eggs. Wireworms

are fond of carrots and potatoes and may be trapped in these in market gardens and glass houses. Attempts have been made to coat the seeds with various substances but they have failed to prevent attacks. In attempting to reduce their numbers it must be remembered that clean farming is extremely helpful as large numbers are starved if the land is free from weeds for any length of time. Clean fallowing will get rid of considerable numbers. It must also be remembered that the grasses in hedges and headlands provide food. A crop of mustard will also starve a great many of them.

In breaking up grass land, leys, or any covered land we must not forget that wireworms are usually present and all crops except mustard are likely to be attacked.

Paring and burning has proved helpful, as also applications of lime and gas lime, especially if the grasses themselves are burned. These methods do not kill the wireworm, but destroy their food. All these methods aim at starving the wireworms. Applications of various kinds are useful for a time, the wireworms are not killed but simply go deep down into the soil and come up again when the trouble is over. Application of rape cake is thought by many to kill them. It has been proved to benefit the crop to which it is applied as wireworms are very fond of it and leave the roots of plants in order to feed on it. In the following season however we shall find that they are still present in the soil ready to feed on any crop except mustard. Rolling the land in order to consolidate it helps the crop, as it prevents wireworms moving from one plant to another as easily as in loose soil. For a similar reason driving sheep over the land when possible is very useful.

Ploughing early in September will kill some of the pupae and may also turn up some wireworms for birds. A number of birds, such as gulls, larks, rooks, starlings, plovers, and pheasants, help to keep the numbers down. The farmer only sees the damage that these birds do; if he saw the food they ate he would probably be led to think that the number of plants they save in many cases outweighs considerably the number which they destroy. Moles also eat wireworms. Stimulating manures applied with the seed help plants considerably against the attacks of wireworms.

Millipedes and centipedes, which are often confused with wireworms, are not the larval stages of insects, but adult animals closely related to insects. They are readily distinguished by the enormous number of their legs. It is important to know the difference between centipedes and millipedes as the former help the farmer by feeding on insects in various stages and also on worms, snails and slugs. Centipedes should not be destroyed. Millipedes are destructive and feed chiefly on roots and other portions of plants which grow underground, such as tubers and bulbs, and also on strawberries. Centipedes have only one pair of legs on each segment whereas millipedes have two pairs on every segment, except the first four (see Fig. 31 *b*).

Millipedes may be trapped by means of a scooped out mangold, or killed by means of poisoned bait.

Turnip Flea Beetles.

Every farmer knows that in a dry, dusty summer his turnip crop is liable to be damaged or even almost entirely destroyed by what he calls the turnip fly or

flea. If we succeed in catching one of these hopping
insects we shall find that it has two pairs of wings
and is therefore not a fly. The outer wings are
hard and protective, and from this we shall decide
that it is a beetle. The name turnip flea beetle is the
most appropriate as it tells us that the insect is a beetle

Fig. 31 *a*. A centipede.

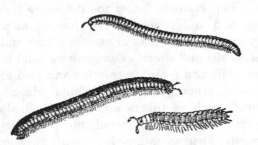

Fig. 31 *b*. Millipedes.

and also at the same time reminds us of its hopping
powers.

In spite of its small size this beetle is capable
of doing enormous damage on account of its numbers
and power of multiplying. Under favourable conditions
it is possible for six generations to hatch out in a
single season. It attacks the plant just at the time

when it is most susceptible, viz. just as it pushes its way through the ground. The adult beetles do the most damage in this case. As soon as the young plant appears they begin to feed on the seed leaves, then on the heart of the plant, and so destroy it entirely. In this way whole fields are eaten bare and have to be resown.

The last brood of beetles formed in the autumn hibernate through the winter in crevices or under leaves. When the warm weather of spring begins, usually at the end of May or beginning of June, they are roused from their slumbers and are prepared to enjoy themselves at the expense of any young turnip plants which may be available.

After a time the beetle lays its eggs usually in the soil near turnips and allied crops or other plants of the same order such as charlock. The small whitish larvae hatch in about 6–11 days and on hatching leave the soil and crawl up the stalk of plants and soon commence to feed on the rough leaves by getting inside them and eating the green portion between the two skins. In this way colourless marks are found on the leaves and by holding them up to the light the young larvae can be seen in them. In about a week they come out and bury themselves in the ground. Here they change into pupae and from these the beetles are formed in about a fortnight (see Fig. 32). These then feed on the turnip leaves, especially of the younger plants when the field has been resown. It is in hot dry weather that the attack is worse, as this favours the development and spread of the beetles. They do not like wet weather and consequently less damage is done in a wet spring. Curtis says they are attracted by the smell of the turnip crop and will fly from considerable distances against the wind. There are several beetles included under the heading of turnip flea beetles but they are similar in their mode of attack. They are very small and measure only about $\frac{1}{10}$ inch in length. Some are of a bluish black colour, the commonest form having a broad yellow stripe down each of its outer wings. Its antennae are one-half the length of its body. By means of its very stout thighs,

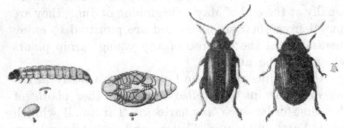

Fig. 32. Turnip flea beetles; egg, larva, pupa and adults. (Magnified.)

especially on the hindmost pair of legs, it is enabled to jump long distances. The larva is yellowish with a distinct head and a pair of legs on each of the first three segments. The hind end of the body is turned down to form a kind of foot.

In addition to turnips, other cruciferous crops, such as swedes, cabbages, and rape, also suffer. Mangold often suffer in the same way from the attacks of a different species: potatoes from another kind and still another species occasionally attacks barley.

Remedial Measures.

The eggs are laid on the under surface of the leaves and do not lend themselves to treatment. As the larvae feed inside the leaves we cannot hope to attack the pest at this stage. It is chiefly against the adult stage that attacks are aimed.

We have seen that the greatest amount of damage is done when the turnips are young, consequently we must aim at helping the plant in these stages and at the same time make things as unpleasant as possible for the beetle.

Drought and frost retard the growth of the plant; the former also favours the beetles and consequently this state of affairs is bad for the turnip.

Clean farming plays a very important part in reducing the numbers of this beetle as weeds, such as charlock, serve to sustain it until the abundant supply of food which the turnip crop supplies is ready. All cruciferous weeds should be destroyed and the grasses in hedges and ditches cut and burnt, as they provide shelter for the beetle during the winter.

The method of cultivating the land is very important, and operations which help the plant in its early stages should be carried on. Cultivation should be done in the autumn if possible as more moisture is retained in the soil and a better tilth obtained. Seeds sown on a fine tilth usually suffer less for this favours the growth of the young plant and is unfavourable to the beetle, as there are few suitable hiding places for it. Careful rolling (if the land is suitable) after drilling the seed breaks up the clods and disturbs the beetles. Miss Ormerod says, "After a long experience I never saw a failure of a braird of turnips on land which had been

long and well prepared before the seed was sown.
Good seed and plenty of it, three pounds per acre,
drilled deeply from (say) two to three inches, ensures
a good start, even in the driest times."

In addition to farmyard manure, artificial manures,
and especially superphosphate, have been shown to be
profitable when sown before or in the drill at a time when
they can help the young plant through the earlier stages.

Sowing mustard with the turnip crop has proved
beneficial in some seasons, but not in others. The
mustard comes up before the turnip and so is attacked
by the beetles. They prefer turnips and usually leave
the mustard as soon as the turnips are ready.

Rolling a piece of infected turnips early in the
morning while the dew is still on the plants is sometimes
practised with success. This process disturbs and
injures the beetle, and also makes the plants dusty,
in which condition they are not so readily eaten.
Driving sheep over turnips while the dew is still on is
also useful for similar reasons.

Many of the beetles can be caught on a wide frame-
work built on wheels with well-tarred boards upon it and
pushed over infested land. The boards are placed at
right angles to each other. The upright ones brush the
plant and disturb the beetles which are usually caught
on the board which projects forward. If the beetles are
very numerous they should be scraped off frequently
and the board kept sticky with tar. Cart grease can
be used instead of tar.

Dragging sacks which have been soaked in paraffin
so that they touch the plants has been tried in many
cases and proved beneficial.

Larks eat the beetles in considerable numbers.

Chafers.

The so-called *white grubs* are the larvae of chafer beetles. There are four kinds which are very destructive to plant life: (1) *Melolontha vulgaris*, the common cockchafer or May bug; (2) *Rhizotrogus solstitialis*, the summer chafer or midsummer dor; (3) *Phyllopertha horticola*, the garden chafer or bracken clock; (4) *Cetonia aurata*, the green rose chafer.

These white grubs often do considerable damage to the roots of plants, causing many to die and others to look very sickly. Grass land is especially liable to attack and whole patches at times die off from the ravages of these grubs. This injury is sometimes attributed to wireworms. Young trees, turnips, cabbages and other farm crops are also injured by attacks on their roots. In addition to the damage caused by the grubs, the beetles themselves feed on the leaves of various trees. The adults are easy to distinguish from one another. They are all characterized by having feelers ending in flattened or leaf-like structures folded one on top of the other.

The cockchafer is the largest and usually measures more than an inch in length. The head and thorax are black, the wing cases brown, each case having five longitudinal smooth ridges. The whole body has a mealy appearance. Five of the abdominal segments have a white triangular marking on each side. The end of the antenna is composed of six leaves in the female and seven in the male (see Fig. 33 a, 1).

The summer chafer measures about ¾ inch in length, is brownish in colour and covered with a quantity of rather long hairs. The markings on the abdomen are not so distinct as in the case of the cockchafer. The end of the antenna has only three leaves (see Fig. 33 a, 3).

Fig. 33 a. (1) *Melolontha vulgaris*, the cockchafer; (2) *Cetonia aurata*, the green rose chafer; (3) *Rhizotrogus solstitialis*, the summer chafer; (4) *Phyllopertha horticola*, the garden chafer.

The garden chafer is not quite ½ inch in length. Its head and thorax are of a metallic green colour, and its wing cases brown (see Fig. 33 a, 4).

The green rose chafer has its thorax and wing covers of a metallic green colour (Fig. 33 *a*, 2).

The grubs of these are not so easy to distinguish and can best be identified by a microscopic examination of one of the jaws[1]. The grub is of a whitish colour. When at rest it lies on its side with its body curled in the form of a semi-circle. It has a distinct head with marked jaws and feelers. The three segments behind

Fig. 33 *b*. *Melolontha vulgaris*, the cockchafer; (5) the larva; (6) the pupa; (1) the beetle.

the head have each a pair of jointed legs. The abdomen is thick and fleshy, and swollen at the tail end which is often darker in colour, the contents of the intestine being visible through the body wall. The body is wrinkled and each segment except the second, third, and last bears a pair of spiracles. When fully grown the grubs vary considerably in size[2].

The grubs of the cockchafer and the green rose chafer measure from $1\frac{1}{4}$–$1\frac{1}{2}$ inches in length. They

[1] See *R.A.S.E.*, VIII. 1897, p. 748.

[2] The grub of the dung beetle is very similar but has a very short third pair of legs.

are easily distinguished by the red spots which the latter possesses on each side of the segment behind the head. When full grown those of the summer chafer measure over an inch in length, whereas those of the garden chafer only measure ¾ inch.

The life histories of these beetles differ chiefly in the length of time from one generation to the next. The cockchafer takes four years to complete its life cycle in this country, but in some countries it may take three or five. The length of life of the summer chafer is not known with certainty, but is thought to be either one or two years. The garden chafer has a new generation every year. The green rose chafer has a new brood every two or three years.

Cockchafers make their appearance in May and are often abundant for five or six weeks. They remain concealed during the day and fly about after dusk. The leaves of oak trees are their favourite diet. The female burrows into the earth to lay her eggs, preferably in loose soil or grass land, and lays about ten eggs. The grubs hatch from these in about six weeks. During the first summer they do very little damage as decaying matter is their chief food. In the following years they feed on plant roots. In winter they escape the unfavourable conditions by going deeper into the soil. After spending three years in the grub stage they scoop out recesses deep down in the soil and change into pupae. The beetles come up in the following May. The cockchafer grubs are responsible for most of the damage done to young trees. Young oaks and conifers suffer considerably.

The life history of the summer chafer is very similar, the beetles appearing after the cockchafers in

June and July. The garden chafer also appears in
June and July, and flies about in bright sunshine.
These chafers sometimes feed on young apples.

The green rose chafer flies about during the day
and appears at the same time as the cockchafer. The
adults are especially harmful to the flowers of roses,
strawberries and turnips.

Remedial Measures.

These grubs feed chiefly on grass land and hence
are extremely difficult to eradicate. Surface application seems to be of little value, as the grubs can descend
out of its reach. It is chiefly against the adult stage
that measures must be taken as they are then more
likely to succeed. On the continent co-operative
attacks are made against the adult chafers which
never travel far to lay their eggs; hence the districts
in which they are destroyed reap the benefit.

In this country the summer chafer and the garden
chafer seem to be the cause of most of the damage.
The cockchafer is very local, but may be abundant in
those districts in which it is found.

The importance of knowing the life history of these
chafers is evident if we are attacking them at the
adult stage. It takes four years for cockchafers to
produce another brood of cockchafers, so that having
noted when the beetles are numerous we know when to
expect the next brood. The date of appearance is also
important. We must be prepared to attack the cockchafer and the green rose chafer in May before they have
laid their eggs, as no advantage is gained by destroying
them afterwards, whereas the summer chafer and the
garden chafer do not appear until June. The small

amount of damage which the cockchafer does during the first year of its larval life often leads the farmer to suppose that the grubs are not present whilst they may be found in the young stage under healthy grass.

The cockchafer and green rose chafer fly at night, and are very sluggish during the day time, when they may be found on trees, usually isolated ones, in fields and hedges. They rarely go into a wood except in very rough weather. The beetles can be beaten from trees and shrubs by means of long sticks, collected on sheets and then destroyed. The best time for this operation is in the morning; the trees must not be shaken too much or the beetles will fly away. In this way a single farmer in France has collected nearly a ton of cockchafers. Pigs will readily eat them if they are shaken on to the ground.

The summer chafer and the garden chafer must be attacked in the early morning, in the evening, or on very dull days, as they fly about in bright weather. On arable land frequent hoeing has been found to be useful and in gardens handpicking of the grubs is profitable. The grubs can be killed by naphthalene (2 cwts per acre). In gardens the larvae may be trapped by placing pieces of turf grass downwards just under the soil.

When badly attacked grass land is ploughed up pigs and chickens should be turned into the field as they will destroy large quantities of the grubs. Rooks, starlings, green plovers, gulls and moles devour a large number of them, and the damage to grass land, which is often attributed to rooks, is really due to the grub, for which they are searching. The chafers themselves are said to be eaten by rooks, owls and nightjars.

Grain Weevil.

Considerable damage is often done to stored grain by the larvae of two small beetles, *Calandra granaria* and *C. oryzae*. They attack all kinds of grain, such as barley, malt, wheat, maize, and oats, and are troublesome to cargoes of grain as well as in granaries and breweries. The trouble begins in the spring and summer. The beetle bores a small hole in the grain by means of its snout or proboscis; into this hole it proceeds to deposit an egg and then closes the hole with saliva. One egg is deposited in each grain and in this way a lot of grains are infected by a single beetle. The larva which hatches from the egg feeds on the grain, devouring all except the skin and the germ. When it has eaten its fill it changes to a pupa and shortly afterwards into the adult beetle. This then eats its way through the husk, leaving a large hole (see Fig. 34 *b*). The time taken to complete the life cycle depends largely on the temperature. Under favourable conditions such as a close atmosphere this may take only a month. It is difficult to detect any damage that is being done by inspecting the grain, as there is little difference between the injured and the uninjured grain. If, however, a quantity of the grain is placed in water the partly eaten grains float on the surface.

The weevils cannot stand cold and on the approach of winter usually migrate to cracks or shelters in the floors and walls. In spring they return once more to any corn which they may find. *C. granaria* is unable to fly, but *C. oryzae* possesses flying wings.

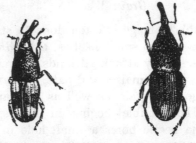

Fig. 34 *a*. Grain weevils. *Calandra oryzae* (on the left), *Calandra granaria* (on the right). (Magnified.)

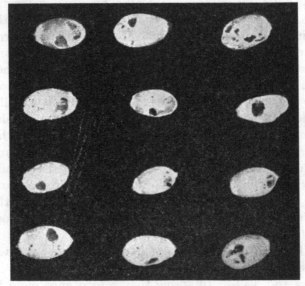

Fig. 34 *b*. Wheat damaged by grain weevils.

The beetle is about ⅛ inch long and of a dark-brown colour. As is the case with all weevils it has a long snout or proboscis, carrying at its base a pair of elbowed feelers (see Fig. 34 a). The mouth is at the end of the proboscis. The larva is a small white legless grub with a wrinkled body, brownish horny head, and distinct biting jaws. These beetles are very prolific, and as only one larva is required to destroy each grain, the introduction of a few weevils may mean a considerable loss.

Remedial Measures.

As the beetle is averse to cold we can do some good by keeping our barns well ventilated so as to keep the temperature as low as possible. By leaving the windows or door of the granary open on frosty nights the beetles are checked. They do not go very deep into a heap so that storing in bulk will keep a considerable quantity of grain free from the beetle.

If we find our grain attacked we have a fairly easy method of killing the pest. This consists in fumigating with carbon disulphide. In a closed bin it is a simple matter and consists in putting the carbon disulphide in saucers on top of the corn. About 1 lb. to 100 bushels is sufficient (the amount depends on the space, a small quantity of corn in a large bin will require more than the above quantity). The bins should be opened after about 24 hours.

If the grain is stored in a barn all holes should be stopped up and the barn made as air-tight as possible. Carbon disulphide is then put on the top of the grain in saucers, 1 lb. for every 1000 cubic feet of space.

If the heap is covered with canvas a smaller quantity

is sufficient and in a large barn this covering is often necessary.

It must be remembered that carbon disulphide is heavier than air and readily descends.

Treatment should begin at the bottom of a building if several floors are being treated. After fumigation the bin or granary should be opened for two or three hours before entering the building as the fumes are poisonous. No naked light should be taken into the treated building as the fumes are highly inflammable. When empty the walls of the granary should be whitewashed, a little carbolic acid being added to the wash. All crevices in the walls should be stopped up to prevent the beetles from hibernating there.

Gall Weevil or Club (Ceutorrhyncus sulcicollis).

This pest may become very troublesome when it has established itself in market gardens where cabbages are grown annually. It attacks other members of the same order such as turnips and swedes, and in these cases indirect damage may be done, from the entrance of organisms which cause the roots to rot. When cabbages are attacked their growth is stunted, the leaves become yellowish, and in many cases the plants are killed. The presence of this pest may be recognized by the swellings on the underground portion of the stem in cabbages, and on the underground succulent part of turnips (see Fig. 35). These swellings give the name of "club" to the disease and sometimes lead to its confusion with finger and toe, which is caused by a fungus (see p. 35), or root knot, which is caused by an eelworm (see p. 170). It is readily distinguished from these by examining the swellings of a diseased plant,

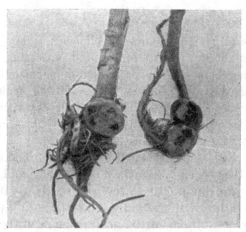

Fig. 35 a. Cabbage plants attacked by the turnip gall weevil
(*Ceutorrhynchus sulcicollis*).

Fig. 35 b. Swede turnip attacked by the turnip gall weevil
(*Ceutorrhynchus sulcicollis*).

which contains small curled yellowish legless grubs measuring about ¼ inch in length and possessing brown jaws. These are the larvae of small blackish beetles possessing the long snouts with elbowed feelers so characteristic of weevils. These weevils lay their eggs on or in the cabbage stalks or turnip roots and from these eggs the young grubs hatch out.

By feeding on the plant these cause the food supply to go to the places where they are feeding and hence swellings are produced. After a time they eat their way out of the swellings and go into the soil to pupate in earth cases. Some of the grubs remain in the plant through the winter, whereas others pass the winter as pupae. In the following spring and summer the weevils come out and eventually lay their eggs on suitable plants.

Remedial Measures.

In ordinary farm practice where cruciferous crops are grown every four years this weevil does little damage as it has great difficulty in living from one crop to the next when cruciferous weeds are kept down. In market gardens these crops occur more frequently and they become infested. The practice of putting the cabbage stalks in a compost heap or ploughing them under is favourable to the insect as the grubs go into the soil, pupate and so give rise to weevils next spring. The stalks containing the grubs should be burnt. Care must be taken in planting to destroy all young plants which have galls on them.

Gas lime in the autumn checks the weevil. Rooks and partridges eat the grubs in the galls which they are able to reach. Cruciferous weeds should be kept down.

The grub of an allied weevil *Ceutorrhyncus assimilis* feeds on turnip seeds while they are still in the pods.

Fig. 36 a. Sitones weevils. *Sitones lineatus* (on the left), *Sitones crinitus* (on the right). (Magnified.)

Fig. 36 b. Broad bean leaves attacked by Sitones weevils.

Pea Weevils (*Sitones lineatus* and *Sitones crinitus*).

These weevils attack peas, beans, and clover in a very characteristic manner by eating semi-circular patches out of the edges of the leaves (see Fig. 36 b).

In this way whole leaves may be eaten. As the weevils are not easily seen the damage is often attributed to birds.

They may be found by carefully examining plants that are attacked. On touching the plant they fall to the ground and remain motionless. The eggs are laid in the soil and the legless larvae feed on the roots of peas, beans, and clover, and often bore channels along the roots. In winter the weevils hibernate.

Remedial Measures.

These pests are difficult to keep in check. A fine tilth is unfavourable to them as their hiding places are destroyed. For clover and field peas, light rolling followed by a good dressing of soot is said to be useful. Poultry eat the weevils readily.

The Pea and Bean Seed Beetles (*Bruchus pisi and Bruchus rufimanus*).

These beetles are very closely allied to the weevils. They have a short snout, but their antennae are not elbowed (see Fig. 37). The larvae resemble weevil larvae, being legless and wrinkled. They feed on the seeds of peas and beans, eating some of the stored food but without injuring the young embryo.

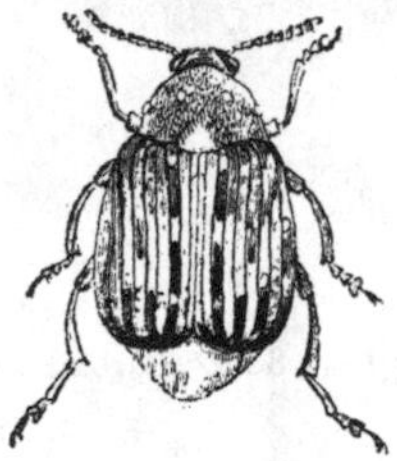

Fig. 37. *Bruchus rufimanus*, the bean seed beetle. (Magnified.)

The eggs are laid on the young pods in the flowering stage, and the larvae which hatch out make their way into the seeds. Each larva pupates in the seed where it has fed and the beetles formed from these also remain in the seed for some time. The seeds in which the beetles occur have a small round pit in the skin of rather different colour from the rest; no holes are made until the beetles escape.

Remedial Measures.

The beetles may be destroyed by fumigating with carbon disulphide as in the case of grain weevils (see p. 121).

When seeds that have been attacked are sown they take longer to germinate and under adverse conditions may not produce healthy plants; under favourable conditions however plants very similar to those from healthy seed are produced, the only difference being that they are delayed a few days in the young stage.

CHAPTER XI

FLIES

The Frit Fly (Oscinis frit).

This pest has for many years caused considerable losses to farmers by its attacks on corn crops.

Spring oats and winter wheat or oats (after leys) suffer most but many of our commoner grasses are attacked as are occasionally barley and rye.

The damage to oats is often confused with attacks of the eelworm *Tylenchus* (see p. 168), but it may readily be distinguished by the yellowing of the central shoot while the older leaves remain green.

In a badly infested field of spring oats very few
ears are formed as the shoots which would normally
bear the ears are killed. Some plants may be killed
but most of them send out fresh tillers which in turn
may be attacked. A badly attacked plant looks like
a tuft of very coarse grass (see Fig. 38).

Winter corn is not attacked until February and
March and then a good crop is thinned out by many
of the plants being killed. Where the wheat is tillering
well some of the shoots of an attacked plant may
escape but on loose soils where the tillering is poor
plants are more easily killed. The attack is similar in
all plants, i.e. the central shoot is killed and turns
yellow whereas for some time the older leaves remain
green. [In the case of wireworm attack the whole
shoot goes yellow and usually several plants in suc-
cession.]

If we carefully examine a plant which is attacked
we shall find a small maggot in the middle of the plant
feeding on the young shoot. This is the larva of the
frit fly, and is about $\frac{1}{8}$ inch long when full grown. It
has a somewhat transparent, shiny, fleshy appearance.
Its body is segmented and legless. The anterior end,
which is somewhat pointed, is easily recognized by a
pair of almost black jaws which are capable of being
protruded. Near the front end on each side is a
branched spiracle. The posterior end is blunt and
possesses two wart-like spiracles (see Fig. 39).

The maggot eats into the stem and in many cases
may eat through it. The plant responds by forming
other shoots and so we get a tufted appearance.
Diseased shoots may be found together with healthy
ones on the same plant, especially in mild attacks.

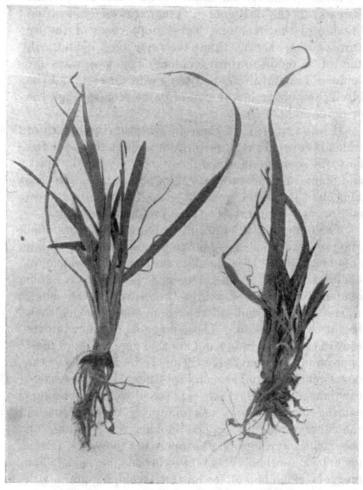

Fig. 38. Oat plants attacked by the frit fly (*Oscinus frit*).

Later in the season small brown cases are present instead of the maggots. These are easier to find because of their colour. These pupa cases of the fly are slightly shorter than the maggots, cylindrical, and of a reddish-brown colour. The two wart-like spiracles are still present at the posterior end, and the black jaws of the larvae can often be seen through the case (see Fig. 39 *B*).

If we keep some of these in a bottle (the mouth of which is covered with muslin), for a short time, the frit flies will soon hatch out. They are very small, black and shiny and measure less than ⅛ inch in length (see Fig. 39 *C*). This is precisely what happens in nature and every year there are three or four generations.

The first generation of flies hatches out in April and May and lay their eggs on the young shoots of spring oats or grasses. It is usually late sown oats which are badly attacked. From these eggs maggots soon hatch out and bore their way into the centre of the shoot, where they feed and kill the young shoot, which then begins to turn yellow. These maggots begin to pupate towards the end of May and another generation of flies[1] appears in June and July. They lay their eggs near the developing ears (and also on late tillers and grasses) and the maggots from these eat the developing flowers causing blind flowers. (As many as 10 per cent. of the flowers may be spoiled in this way.) Later on these flies lay eggs on the open oat flowers. (Each generation probably lays two broods of eggs.) The maggots which hatch feed on the developing grain with the result that a number of them are empty. (In 1922 a sample sent in contained nearly 60 per cent. of empty grains.) Occasionally the whole of the top of the plant

---

[1] The first generation of flies die before or soon after the appearance of the second generation.

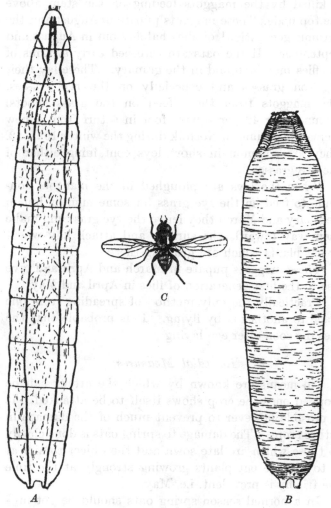

Fig. 39. The frit fly (*Oscinus frit*). *A*, the larva; *B*, the pupa; *C*, the fly. (Magnified.)

is killed by the maggots feeding on the stem above the top node. These maggots pupate in August and the autumn generation[1] of flies hatches out in August and September. If the oats are threshed early swarms of the flies may be found in the granary. These lay their eggs on grasses and especially on the rye grasses. The maggots from these feed on the grass shoots. I found over 40 per square foot in a turf from a new recreation ground in Norfolk during the winter 1921–22. They are common in short leys containing either of the rye grasses.

When the leys are ploughed in the maggots continue to feed on the rye grass for some months but in February and March they leave the rye grass and make their way towards the surface and attack any wheat or oat plants which are growing there.

These maggots pupate in March and April and give rise to the first generation of flies in April and May.

Practically the only method of spreading from one place to another is by flying. It is probable that the flies die soon after egg laying.

Remedial Measures.

No means are known by which the attack can be stopped once the crop shows itself to be attacked. It is possible however to prevent much of the damage in future crops. The damage to spring oats is done chiefly to those which are late sown and the object to aim at is to get the oat plants growing strongly at the time the frit fly is prevalent, i.e. May.

In a normal season spring oats should be got in— if possible—by the middle of March on a good tilth. (Sowing early on a bad tilth may be more disastrous

---

[1] There are possibly four generations produced in one year.

than sowing later.) If driven later than this it will probably be more profitable to sow another crop. In 1912 a field of oats in Norfolk sown the second week in April was ruined by frit fly, whereas a neighbouring field sown the second week in March was almost free from it. In 1922 which is the worst frit fly year I have ever known—early sown oats suffered considerably but not so badly as late sown ones. This was probably due to the cold conditions which prevailed when the oats were sown and which gave them a severe check. Oats which tiller freely suffer less than the poor tillerers such as Black Tartarian.

Quick-acting manures such as nitrate of soda and superphosphate will help the oats through the early stages if the soil is not in good heart.

Winter wheat (or oats) is only attacked by frit fly after a ley containing grasses and consequently we have a choice of the following methods to prevent the damage done by this generation of maggots.

(1) Plough or broadshare the grass leys before harvest, i.e. before the frit fly lays its eggs on the grasses. It is necessary that all the grasses should be buried or killed. Spring corn may be sown after a ley as the plants are not available in March when the maggots require them.

(2) Leave rye grasses (and other grasses) out of the leys which are to be followed by wheat or oats. Any leguminous plants can be used.

(3) Change the rotation so that winter corn does not immediately follow a ley, e.g. roots—barley—seeds —potatoes—winter corn.

This pest does much more damage on a loose tilth than on a firm one. Consequently it is important to

get a good firm seed-bed in order to get the plants to
tiller well and so escape part of the damage.

Gout Fly (*Chlorops taeniopus*).

This fly attacks barley, wheat, rye, and certain
grasses. In this country the injury to the barley crop
is often very serious.

In a badly attacked plant the shoots are swollen
considerably suggesting the name "gout." The ears
are usually unable to force their way through the
leaves which become spirally twisted at the top (see
Fig. 40). A groove is eaten out along the upper
part of the stem and the lower grains are usually
shrivelled. The plant itself is stunted and distorted.
On cutting open a diseased plant at the earliest stage
of an attack a small maggot may be found feeding on
the upper portion of the stem. This larva of the gout
fly is somewhat similar to the frit fly maggot, only
considerably larger. One maggot only is usually
present in each shoot. When fully grown it measures
from $\frac{1}{4}$—$\frac{3}{10}$ inch in length (see Fig. 41 A). It has a semi-
transparent, shining, fleshy appearance and is yellowish
in colour. It possesses no legs and is segmented.
The contents of the gut can be seen through the body
wall. At the pointed anterior end are a pair of small
black jaws, and at the posterior end a pair of spiracles
protrude. If we examine a diseased plant later in the
season we shall find that the maggots have changed
into pupa cases. The pupa case is slightly smaller and
of a brownish colour (see Fig. 41 B).

From these the flies hatch out in August and
September. They may be hatched out by keeping
puparia in a bottle covered with muslin. They are

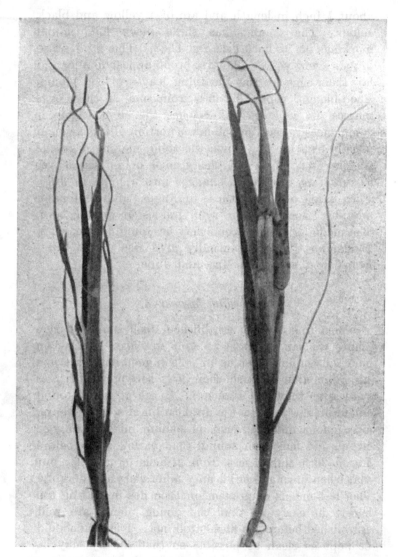

Fig. 40. Barley plants attacked by the gout fly (*Chlorops taeniopus*).

about $\frac{1}{6}$ inch in length and are of a yellow and black colour. The thorax has three black longitudinal markings on its back (see Fig. 41 *C*). This fly has two broods every year. The first brood appear in May and lay their eggs on the sheathing leaves of the plants. The maggot which hatches from one of these eggs pierces its way until it reaches the stem. Here it commences to feed on the lower portion of the ear, and usually eats its way down the stem, making a groove to the first joint. In the groove or in the leaves surrounding the ear it changes into a pupa in July. Flies hatch out from these in August and September and lay their eggs on wild grasses or winter corn. The maggots of this brood may be found on grasses in November. They eventually give rise to the first brood next season in May and June.

Remedial Measures.

Once this fly has established itself in our barley plants we can do little to stop any loss. Barley on poor soils suffers most, as on soils in good heart the ears are often able to push their way through the leaves even after they are attacked. If the attack is found out in its early stages the application of a quick-acting nitrogenous manure, such as sodium nitrate, may be useful. It has been found that barley sown before March 20th suffers less from attacks of this fly, but that when sown in April it may suffer very considerably. The best means of prevention then lies in getting the barley in early so that the young plants are well established before the flies hatch out. The land should be kept as clean as possible so that the fly may be

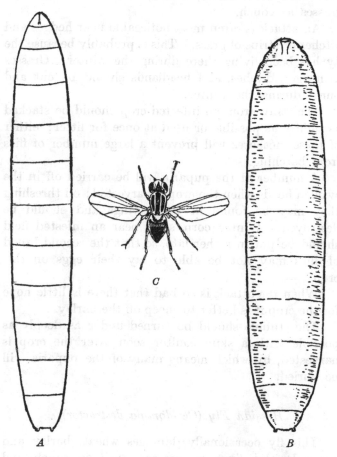

Fig. 41. The gout fly (*Chlorops taeniopus*). *A*, the larva;
B, the pupa; *C*, the fly. (Magnified.)

prevented from living during the winter on such grasses as couch.

An attack is often most noticeable near hedges and ditches or strips of grass. This is probably because the fly has been living there during the winter. Grasses in hedges, ditches, and headlands should be cut and burnt during the winter.

The straw from an infected crop should be stacked as tightly as possible or used at once for litter; either of these measures will prevent a large number of flies from hatching.

A number of the puparia will be carried off in the swollen heads when the crop is harvested; on threshing they may be found in the cavings and should be destroyed. Winter corn sown near an infested field should be put in rather late, so that the second brood of flies may not be able to lay their eggs on this crop.

When the attack is so bad that there is little hope for the crop it is better to sheep off the barley.

The stubble should be turned under as deeply as possible with a skim coulter soon after the crop is harvested, by which means many of the puparia will be buried.

Hessian Fly (*Cecidomyia destructor*).

This fly occasionally damages wheat, barley and rye. It also lives on grasses, such as couch and timothy.

Wheat suffers most from it and the effect is usually noticed when in full ear. Plants here and there are seen to have fallen down. This condition goes by the

name of dog-leg, elbowing or "strailed corn." On one
of these shoots may be found below the elbowing, just
above one of the lower nodes and underneath the leaf
sheath, one or more maggots or pupa cases, usually the
latter.   The maggot is a small whitish, legless grub and
is easily recognized by a curious process just behind
the head, known as the breast bone.   The pupa cases
found in the same position as the larvae are usually
known as *flax seeds*, which they somewhat resemble,
but they are considerably narrower.   They are $\frac{1}{8}-\frac{3}{16}$ inch
in length and of a brownish colour.   When the wheat
is cut some of the pupae are carried away in the straw
and others remain in the stubble.   Some of them give
rise to flies in autumn which lay their eggs on such
grasses as couch and timothy.   In America this brood
does enormous damage to autumn sown wheat, but in
this country it gives little trouble.   Most of the pupae
hatch out in spring and the eggs are laid on the growing
wheat plants.   The flies are gnat-like with bodies
about $\frac{1}{8}$ inch in length and with long legs and feelers.
This pest is preyed on by a number of parasites, and
large numbers of these come out of the so-called flax
seeds.

Remedial Measures.

A stiff strawed wheat does not go down under an
attack nearly so much as a weaker strawed variety.
In one case where a crop was being attacked two kinds
of wheat had been sown in the same field, the weaker
strawed variety was badly laid but in the stiffer
strawed form it was difficult to find an elbowed stem.

An infected stubble should be deeply ploughed in
the autumn.   Complete inversion of the soil by means

of a skim coulter buries the pupae, and so prevents them giving rise to flies in the following spring.

The screenings of infected wheat often contain "flax seeds" which should be destroyed.

The grain of an infected crop should be carefully examined if used for seed. Couch should be kept down as much as possible.

Wheat Bulb Fly (*Leptohylemyia coarctata*).

This fly has been the cause of considerable loss to wheat growers. It is much more prevalent in some seasons than in others. The worst attacks follow a bare fallow or bastard fallow but bad attacks also occur after crops of potatoes, rape, swedes, turnips and mangolds, especially where the soil is bare during the summer, as would obtain with poor crops of roots or where potatoes are dug early or the tops die off early in the season.

In addition to wheat, winter barley can also be attacked. An attacked plant resembles one which is attacked by frit fly in that the central shoot is killed and turns yellow while the older leaves remain green.

(For life-history see Appendix.)

Cabbage Root Flies.

There are several flies whose maggots attack the roots of cabbage and allied plants, but a description of the cabbage root fly (*Chortophila brassicae*), which is the commonest, will serve to show the method of attack and how to deal with it. This fly attacks all cruciferous crops, but the greatest damage is done to cabbages, cauliflowers and turnips.

When a cabbage plant is attacked the outer leaves turn yellow, the plant makes little growth and soon

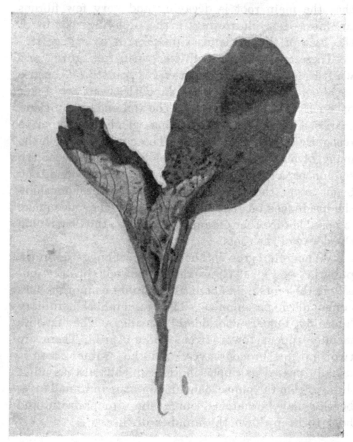

Fig. 42. A cabbage plant which has been attacked by the cabbage root fly (*Chortophila brassicae*). The maggot is shown at the side.

withers. On pulling up one of these plants we find that the main root is decaying and very few fibrous roots are left; burrowing in the decaying root or at the base of the stem are a number of fly maggots.

These are about $\frac{1}{3}$ inch long when full grown, and whitish in colour. The front end is pointed and carries a pair of hook-like jaws. Behind the head are a pair of much branched spiracles. The tail end is blunt and carries a number of projections which enable us to recognize it (see Fig. 43 *A*). The eggs from which the maggots hatch are laid just below the surface of the soil, on or very close to the plants. On hatching they begin to eat the outer layer of the root and reaching the inside feed on the softer portions. They also make it possible for other organisms to enter, thus hastening the decay of the roots.

When fully grown the maggots change to brown puparia (see Fig. 43 *B*) usually in the soil, but occasionally in the roots. In about a fortnight or more the flies come out of these pupae. They resemble the ordinary house fly, but are considerably smaller. The first flies usually appear towards the end of April. There are two or three broods a year and the winter season is usually passed as pupae in the soil, but also as adult flies. From the pupae ichneumon wasps and small rove-beetles may be hatched out; these are parasitic and help to keep down the number of flies.

Remedial Measures.

In America these flies have been successfully kept in check by preventing them from laying their eggs in the soil near the cabbage roots. This is done by means of tarred cards about 3 inches across. They

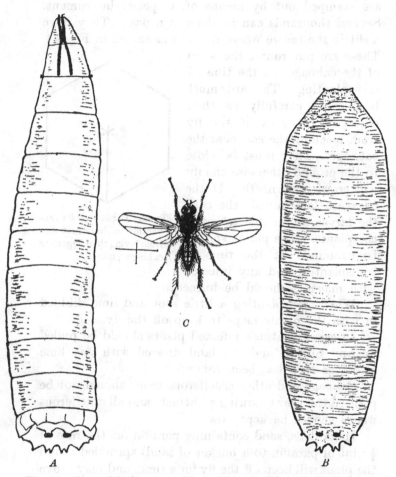

Fig. 43. *Chortophila brassicae.* The cabbage root fly. *A*, the larva;
B, the pupa; *C*, the fly. (Magnified.)

are stamped out by means of a special instrument. Several thousands can be done in a day. They have a slit to the centre where it is cut as shown in Fig. 44.

These are put round the stem of the cabbages at the time of transplanting. The cards must be put on carefully as their object is to prevent the fly from reaching the soil near the cabbage. They must be close to the ground otherwise the fly will crawl underneath. If the eggs are laid outside the card the larvae are unable to reach the plants. The plants should be examined at the time of transplanting and any infested with maggots should be burned.

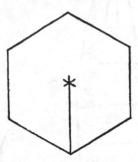

Fig. 44. Cards used for preventing cabbage root flies from laying their eggs near cabbage plants.

When transplanting a little soot and lime dibbled in with the plants helps to keep off the fly.

In cases of attack infected plants should be pulled up and burned and the land dressed with gas lime after the crop has been taken.

Cabbage and other cruciferous crops should not be planted the year following an attack, and all cruciferous weeds should be kept down.

In gardens, sand containing paraffin (at the rate of $\frac{1}{2}$ pint of paraffin to a bucket of sand) sprinkled round the plant will keep off the fly for a time, and may prove useful if done once a week until the plants have got a good start.

The puparia remain in the soil or in the old stumps and it is important that means should be taken to prevent them from hatching. All diseased stumps should be burned and the soil should be turned under as deeply as possible as soon as the crop is removed. It is chiefly in gardens where cruciferous crops are grown every year that this pest proves most destructive. When the crops are rotated the fly has great difficulty in living from one crop to the next.

The carrot fly (*Psila rosae*) and the onion fly (*Hylemyia antiqua*) have life histories very similar to the cabbage fly.

The damage done by the former may be reduced by keeping the carrots earthed up as tightly as possible.

Experiments are at present being carried out to find suitable methods for checking these pests. All onions attacked together with the maggots should be destroyed.

Leather Jackets (*Tipula sp.*).

Leather jackets are the grubs of the flies which are commonly known as daddy longlegs or crane flies. There are several different species which injure crops, all very similar in appearance and in the damage which they do.

It is the larvae or grubs which feed on nearly all our crops and often do enormous damage to grass land and roots. Clover leys and corn may also suffer considerably. These grubs are of a dirty or smoky colour and somewhat difficult to see in the soil. They vary from $\frac{1}{2}$–$1\frac{1}{4}$ inches in length, the longer ones being about $\frac{1}{6}$ inch across, the others correspondingly thinner. They are legless but possess a definite retractile head with a pair of jaws. The tail end is thick and carries six conical papillae. Their skin is tough and wrinkled,

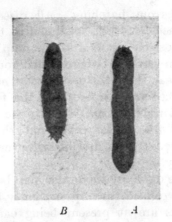

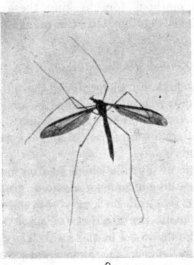

Fig. 45. *Tipula* sp. Daddy longlegs. *A*, the larva or leather
jacket; *B*, the pupa; *C*, the daddy longlegs fly.

hence their name Leather Jackets. The pupa is about the same size as the grub from which it is formed. Its abdomen is covered with a number of spines. Its head bears two horn-like projections (see Fig. 45 *B*). When the adult fly is about to come out the pupa moves to the surface so that its upper half is exposed. The empty cases are often a nuisance on putting greens.

The flies themselves are too well known to need description. According to Curtis the eggs are laid as the insects fly or when they are at rest amongst the herbage and are propelled as from a pop-gun. This takes place in damp or shady places, preferably on grass land or on the grasses of ditches and hedges, chiefly in late summer and autumn, but in some species they may be laid in spring. The grubs live for about eight months chiefly on various roots. They may however feed above ground at night. In some species there are two broods, the first brood of flies appearing in the spring and the second in the autumn. A large number of the grubs are often found after a wet autumn.

Remedial Measures.

All wet land should be drained if possible and tufts of rank grasses should be kept cut. The grasses of hedges and ditches should also be cut and burnt.

Bush harrowing or rolling pastures when the flies are seen in large numbers will kill some of them and prevent others from laying eggs.

Pasture land and leys should be broken up early in autumn if possible. A large number of substances have been applied to grass land in order to try and kill

the grubs, but being very tenacious of life they are not
killed unless the application is so severe that the grass
is also killed. Heavy rolling early in the morning
when the grubs may be on the surface kills some of
them, and also consolidates the soil so that others
cannot move about so freely. This is particularly
useful when corn crops are attacked. They are eaten in

Fig. 46. Wasps eating daddy longlegs flies. These wasps
were destroyed in the nest.

large numbers by rooks, starlings, plovers, pheasants,
gulls, and moles, and to a smaller extent by thrushes
and blackbirds. Considerable numbers of adults may
often be found in wasp nests (see Fig. 46), and they are
also eaten by flycatchers, rooks, swallows, and spar-
rows. When a root crop is attacked frequent hoeing
is useful as many grubs are killed and others are turned
up and eaten by birds. The penning of sheep on the
land when the pupa cases begin to force their way up
to the surface has proved successful.

Wheat Midge (*Cecidomyia tritici*).

This pest damages wheat by injuring the grain and causing it to become very shrivelled. In an ear that is attacked we shall find inside the chaff a number of small yellowish or orange coloured legless grubs about $\frac{1}{12}$ inch long. These suck the juice of the grain. Just before harvest some of the grubs go down into the soil and pupate, forming orange coloured pupae. Others remain in the ears and are harvested; on threshing these may be found in the screenings.

The flies come out about June and lay their eggs usually in the evening on the developing flowers. The flies have yellowish bodies and very noticeable black eyes.

Remedial Measures.

Deep ploughing has proved effective in checking this pest, as the delicate fly is unable to make its way through the soil. All infected screenings should be burnt.

Warble Fly.

The well-known warbles on cattle are caused by the maggots of flies known as *Hypoderma lineata* and *Hypoderma bovis*. The affected cattle have lumps or abscesses on their backs. Each of these lumps contains the maggot of a warble fly.

The life history of the warble fly has not yet been worked out in all its stages. It was formerly supposed that the fly laid its eggs on or in the backs of cattle, and that the maggot bored its way into the skin or was hatched out in it. It has since been shown that the eggs are not laid on the backs where the warbles are found, but are laid on the hairs of the legs and sides of the cattle. They are very difficult to find as they are placed on the hairs close to the skin. The small larvae which hatch out from the eggs have been seen to bore into the skin when placed on it. The next stage of the larva is found in the wall of the gullet and from this it was for some time supposed that the eggs or maggots were licked into the mouth. *How the larvae get from the egg into the gullet is not yet known.*

From the wall of the gullet it is thought that the larva bores its way through the tissues to the position in which we find it under the skin of the back. The eggs are laid during the summer, the actual date varying according to the climate. Flies have been seen as early as June and as late as the middle of September.

The maggots live in the backs of the cattle until May or June. The lumps containing them have holes which ensure a sufficient supply of air.

When fully grown the maggot is about 1 inch long and of a blackish-brown colour. It is segmented and wrinkled. It has no legs but possesses a number of very small hooks which are situated on transverse ridges. The anterior end has a pair of fork-like jaws and at the posterior end are a pair of dark brown spiracles which touch each other (see Fig. 47).

They set up inflammation in the flesh and cause a quantity of pus to be formed, on which they feed. As we should expect from the presence of the spiracles at the tail end, they live head downwards. When fully grown the maggots make their way out of the lumps

A

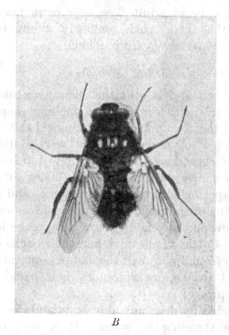

B

Fig. 47. *Hypoderma bovis.* The warble fly. *A*, the larva or warble taken from a bullock's back; *B*, the warble fly. (The fly is only half the length of the photograph.)

and fall to the ground, where they pupate and form almost black pupa cases. This takes place in the early part of the summer, and about a month later the flies come out. They are very like small humble bees. They fly about on hot days and produce a low humming noise which terrifies the cattle and causes them to rush about wildly. On tanning the hides of warbled cattle the skins are found to have holes in them, each hole being caused by a maggot. The surface of the meat of infected cattle has jelly-like patches on it and is known to butchers as "licked beef." These parts have to be cut away.

The presence of adult flies near cows may cause injury as the cattle rush aimlessly about in their endeavour to get away from them.

Remedial Measures.

It has been and is still the custom to smear the backs of cattle with some nasty smelling preparation just before the fly appears in summer; this is done with the view of preventing the warble flies from laying their eggs there. This method of treatment is founded on a false account of the life history of the fly, and such experiments as have been made tend to show that smearing of the backs does little good as we should expect, knowing that the eggs are laid on the legs and shoulders. Cattle which have had their backs covered with cloths from June until the end of September have been found to have warbles on their backs in the following spring.

The warble fly does not like shade and does not go near cattle standing in water. It is very essential that cattle should be provided with a necessary amount of shade so that they can take protection from the fly.

The best method of keeping down this pest is to examine cattle in the spring and to squeeze out the warbles between the thumbs and destroy them. This is a much more satisfactory method than putting some kind of grease or ointment on the animals' backs in the spring in order to stop up the holes. No dressing has yet been found which will kill the maggots without injuring the skin.

The practice of squeezing out the maggots if carried out systematically by farmers would considerably reduce the number of flies, because, as far as we know, the backs of cattle contain nearly all the warbles during the winter, and if these were destroyed few flies would remain to infect cattle the following year. Young stock are more frequently warbled than older ones.

Horse Bot (*Gastrophilus equi*).

Damage is sometimes caused by the maggots or bots of a fly known as *Gastrophilus equi* living in the stomach of horses. When present in large numbers the wall of the stomach may become ruptured with fatal results. In small numbers little damage is done.

The brown and grey fly lays its eggs on the hairs of horses' legs or in some position in which the horse can reach them with its tongue. They are fixed very tightly on the hairs so that the horse cannot lick them off. One fly is said to lay over 500 eggs. In about four or five days the larvae hatch out inside the eggs, where they remain until warmed by the licking of a horse's tongue. They then come out of the egg and pass into the horse's mouth and thence into the stomach. They do not necessarily get into the horse on which the eggs are laid, as another horse may lick the eggs and swallow the maggots. In the stomach the young larvae attach themselves to the wall by means of a pair of strong hook-like jaws. They are enabled to move through the contents of the stomach by means of transverse rows of curved spines pointing backward situated at the juncture of the segments except at the posterior end. The bots feed on the products of inflammation which they set up. They may remain fastened to the stomach wall for nearly a year when they drop off and pass to the exterior in the faeces. When fully grown they may reach about 1 inch in length. They are somewhat pointed in front, but end bluntly at the hind end which carries a large compound spiracle. This can be closed by means of a pair of fleshy lobes which prevent the bots from becoming choked.

If they fall on suitable ground they burrow into it to change into puparia. This usually takes place in June and the flies come out a few weeks later. There are several different species of bots in addition to the one described above, one of which attaches itself to the anus of the horse. Bots have also been found in the brain and spinal cord.

Remedial Measures.

Like the warble fly this fly does not like shade and hence plenty of shelter should be provided in the fields. Well groomed horses rarely suffer from bots, as the eggs are brushed off.

Horses should be watched for the eggs during the months they are likely to be present, i.e. from July to October. The legs should be well brushed or washed with warm water in order to hatch out the eggs. The droppings of infected horses should also be watched for the bots.

Sheep Nasal Fly (*Oestrus ovis*).

This grey fly may be seen in early summer flying around sheep, especially in hot weather. The flies settle on their noses from time to time in order to lay their eggs or to deposit newly hatched larvae. The sheep often try to keep them off by rubbing their noses on the ground and may show uneasiness when the flies are near them.

The maggots pass into the nasal cavities of the sheep where they live for about 10 months on the mucous discharge caused by their presence. Badly infested sheep have a mucous discharge running from their noses, and often sneeze violently. They also rub their noses on the ground. They may frequently shake their heads and hold them on one side. This state of affairs is known as *false gid*, and is due to the presence of the maggots in the nasal passages. It must not be confounded with true gid, which is caused by the cyst of a tapeworm.

When fully grown the maggots come out of the nose or are forced out by the sneezing of the sheep. They make their way just below the surface of the ground and change into brown puparia. From these the flies come out in June and July.

The condition of infected sheep is considerably reduced and in severe cases they may die.

Remedial Measures.

The best method is to try and prevent the flies from laying their eggs on the sheep's nose by covering it with a strong smelling substance such as tar.

A method has been suggested in which the sheep smear themselves. They are supplied with a box of salt to which they have access through holes, about 2 inches in diameter. These holes are then painted over with tar and in this way when the sheep lick the salt their noses become covered with the tar or preparation used. When known to be infested they should be isolated about the middle of April so that the maggots which are sneezed out may be destroyed.

A pasture which has become infested from sheep containing the maggots should be kept free from sheep during June and July.

Sheep Maggot.

The small maggots found on sheep are the larvae of two green bottles, *Lucilia sericata* and *Lucilia caesar*, and the larger ones are those of the blue bottle (*Calliphora erythrocephala*). The smaller ones do the most damage as they dig deeper into the flesh. Lambs suffer more than older sheep, being attacked in the parts around the tail, on the shoulders, and also on the poll. Hot and damp weather seems to favour the attack.

The affected sheep can usually be picked out by the continual wagging of their tails. They also rub and bite themselves to try and get rid of the irritation.

The wool soon becomes discoloured and matted together, in bad cases it falls out and large sores are produced. The affected sheep lose condition very rapidly. The flies lay their eggs from May until October on the positions mentioned, and especially where dirt has collected. Each fly is capable of laying several hundred eggs.

The small legless maggots which hatch out feed first on the surface and then in the sores formed. In about a fortnight they fall to the ground and change into puparia. In another fortnight or more the flies hatch out. In this way several broods are produced throughout the year.

Remedial Measures.

It is the shepherd's business to watch the sheep carefully so that those affected may be treated early. The hindquarters should be kept as clean as possible by docking and clipping the wool of the tail and the regions around. The fly prefers putrid matter in which to lay its eggs so that clean sheep may suffer less. Clean sheep however are often attacked.

Dipping sheep in some poisonous substance is compulsory in this country between certain dates. This order was brought in to reduce scab (caused by a mite). It also kills maggots, keds, lice and ticks. Sheep which have been dipped in this way may become struck by maggot about a fortnight later. In bad cases it may be necessary to dip for maggot once or twice.

A dip composed of sulphur, arsenic, soda, and soft soap is useful for keeping off flies. Dressing the affected parts with a mixture of equal parts of turpentine and rape oil, or by means of paraffin mixed with water, kills the maggots. Dusting sulphur on this serves to keep the fly off for a short time.

A favourite spot for maggots is behind the horns of rams. They may be kept off by means of olive oil and pitch oil dusted over with sulphur.

CHAPTER XII

APHIDES AND SAWFLIES

There are a large number of different kinds of Aphides, or, as they are usually called, Plant Lice, or Green Fly or Blight, and in favourable circumstances, such as a hot dry summer, they are present in enormous numbers. They attack practically all cultivated plants —usually the leaves and stems but sometimes also the roots. They belong to the same group of insects as bugs, and live in a similar way by sucking the juices of plants after piercing them with their mouth appendages, which are well suited for the purpose. They differ considerably from the majority of insects in their methods of reproduction. They have no complete metamorphosis in which there is a resting stage, and only in special cases do they lay eggs.

The life history of a typical aphis will help us to understand them.

We will start with those found in spring. These are without wings (see Fig. 48) and are all females. Each one gives birth to a varying number of young.

At this time of the year no eggs are laid. Instead changes go on inside the mother who gives birth to living young. These begin to feed on the juices of the plants and in less than a week they may also give birth to young. In this way a large number of broods are produced during a season, indeed one green fly is capable of producing several millions in a very short time.

During the summer we find winged forms together with the wingless ones (see Fig. 48).

The winged forms are also females, no males

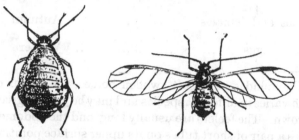

Fig. 48. Wingless and winged Aphides. (Magnified.)

being produced at this time of the year. They also give rise to living young and serve to spread the aphides from one plant to another as one host plant soon becomes overcrowded. It is only at the end of the season that males are produced. They may be either winged or wingless and are born together with the females. Pairing takes place and the fertilized females do not produce living young but lay eggs. These do not hatch out at once but are capable of resisting the cold weather of winter; in spring they give rise to the wingless females who give birth to

living young as described above.　In some cases the
females are capable of living through the winter.

　Their life history may be shortly represented thus:

↓
Wingless females　　...　　..　　...
↓
Several broods of wingless females ...
↓
Winged and wingless females　　　...
↓
Several broods of winged and wingless
　　females　　...　　...　　...　　...
↓
}　Spring
and
Summer

Males and females　　...　　...　　...　Autumn
↓
Eggs　　...　　...　　...　　..　Winter

　The aphides have soft swollen bodies, the colour of
which varies in different species and may be yellow, green
or brown.　The feelers are usually long, and the abdomen
carries a pair of short tubes on its upper surface pointing
backwards and slightly outwards.　These are characteristic of aphides and are called "honey tubes"
because of the sugary liquid they secrete.　Aphides are
covered with a waxy bloom which causes water to run off
them.　The winged forms have two pairs of membranous
wings of which the front pair is much larger than the
other pair and also considerably longer than the body
of the insect.　The males are smaller than the females.
They all live by sucking the juices of plants and have
long probosces for piercing and sucking.　They often
cause the leaves to curl in such a manner that the
insects themselves are hidden away inside the curled
leaf.　They also excrete a sticky substance known as

"honey dew," which may be found on the leaves of trees. It blocks up the stomata and sometimes becomes so dirty, owing to the growth of a saprophytic fungus, as to appear like a coating of soot.

If the aphides were not checked by some natural means it is obvious from the rate at which they reproduce that they would soon cover all the plants on the earth. They have a large number of natural enemies, and nature has given them this capacity for reproduction in order to withstand these. After a bad

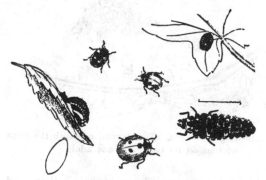

Fig. 49. Lady-birds. An egg (highly magnified), a larva, two pupae and three beetles.

attack of aphis we often find a large number of lady-birds on plants. These small beetles and their larvae devour a large number of aphides. The larvae of the lace-wing flies, also known as "niggers," feed on plant lice and may often be found with the skins of their victims on their backs (see Fig. 50). They are usually difficult to find when at rest.

The legless larvae of hover flies also eat a considerable number of green flies.

On a plant attacked by aphis may be found among the living forms a number of inflated skins of a whitish colour looking like small balls. These are the dead lice which have been killed by ichneumon wasps. These little wasps lay their eggs in the bodies of the lice and as the louse grows its parasite also grows and kills it. From the inflated skins the ichneumon wasps hatch out and continue their useful work of keeping down the number of aphides by laying their eggs in them.

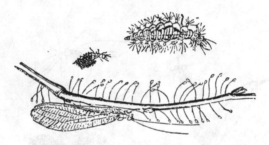

Fig. 50. Lace-wing fly. Eggs. Two larvae, one with the skins of its victims on its back and one adult.

Earwigs eat large numbers of green flies and also their eggs.

Considerable damage is done by aphides in hot dry weather. Rainy weather is unfavourable to their development.

Remedial Measures.

It must be borne in mind that a certain number of green flies are killed by other insects. The best means of keeping down these pests is to spray with some solution containing soft soap as a basis. The soft soap

makes the solution spread more easily over the plant.
Nicotine is the best insecticide to add.

Mixture I.	Soft soap	...	5–10 lbs.
	Water ...	...	100 gallons.
Mixture II.	Soft soap	...	5 lbs.
	Nicotine	...	5–7 ounces.
	Water ...	...	100 gallons.

This is the most effective mixture and is economical
in spite of its extra cost.

Mixture III.	Soft soap	...	10 lbs.
	Paraffin oil	...	1 gallon.
	Water ...	...	100 gallons.

Dissolve the soft soap in a small quantity of hot water,
squirt in the paraffin with a garden syringe and thoroughly emulsify with the syringe after dilution.   These
sprays should be applied through a coarse nozzle.

This spraying is in general use for fruit trees, hops,
and roses.   It is important to spray early just after
the flies have hatched, as little good is done once the
leaves become curled.

In the case of the bean aphis the best method is
to "top" the beans as soon as the flies are noticed.
Spraying with soft soap has also been shown to be
profitable in bad seasons.   This aphis also lives on
docks and thistles, and migrates from these plants to
the bean.

The hop aphis migrates to plums and damsons at
certain seasons and consequently when grown near they
should be sprayed as well as the hops.

The woolly aphis or American blight (*Schizoneura
lanigera*), which is found chiefly on the apple, lives

11—2

also on the elm. Eggs are laid on the elm in late autumn and in the following spring they curl the elm leaves. In July they migrate to apple trees; some however live entirely on the apple and of these some live entirely above ground, others entirely on the roots and others on both shoots and roots.

Painting the affected parts with petrol or paraffin or spraying with soft soap and nicotine is the best remedy.

Carelessness among certain nurserymen in allowing their young stock to become infected is responsible for a large amount of damage done by this pest.

Various winter washes have proved effective in reducing fruit aphides by preventing the eggs from hatching.

Sawflies.

Sawflies are not true flies as they possess two pairs of wings. They belong to the same family as bees and wasps.

Corn Sawfly (Cephus pygmaeus).

This pest rarely does much damage in this country. It attacks wheat and oats.

The eggs are laid in the stems of young plants and the larvae which hatch out feed on the soft substance inside and eat their way through the knots. About harvest time the maggots are fully grown and they make their way to the bottom of the stem where they eat a ring almost round it. The attacked plants may be picked out as the shoots die and the dead ears are easily seen among the ripening corn.

Unlike the larvae of most sawflies the maggots seem to be legless as their legs are too small to be seen.

The maggots are small, curled, wrinkled grubs with a definite head. They pupate in the stem just below the cut, and here they remain until the following May, when they hatch into sawflies.

Remedial Measures.

As the pupae are present in the stubble after harvest, means should be taken to destroy them. The stubble should be burnt if possible : if not it should be harrowed and the plant remains collected and burned.

The larvae of other sawflies may be found feeding on fruit trees. They resemble small caterpillars, but may be distinguished from them by the number of their legs. In addition to three pairs of true legs they possess more than five pairs of false legs, which is a greater number than is found in any caterpillars of moths and butterflies.

CHAPTER XIII

EELWORMS

Eelworms are not insects but true worms. Some of them cause plant diseases and others are present in a large number of diseased plants. They are unsegmented worm-like creatures, very small, and never reach $\frac{1}{4}$ inch in length, the majority being about $\frac{1}{20}$ inch long. For our purpose we may classify them as : parasitic forms; free-living forms.

The former cause plant diseases. The latter are always present in decaying roots or stems, and are regarded as living saprophytically on decaying matter, but it is possible that they hasten the destruction of the plant. Parasites may readily be distinguished from free-living forms, by the possession of a retractile spine-like process at the head end which is used for penetrating plants (see Fig. 51 *C*).

The parasitic forms are comparatively much thinner. They are very sluggish in their movements as compared with most free-living forms. Many of the latter have long tails which help them to move quickly.

Tulip-rooted Oats.

Tulip root in oats is due to two causes, one of which, frit fly, has already been studied. The other cause is due to an eelworm known as *Tylenchus dipsaci* (formerly *devastatrix*). Sometimes both are found in the same plant. The two diseases are very similar in appearance (cf. Figs. 38 and 52). On examining diseased plants containing frit fly we usually find (not in the earliest stage) in addition to the frit fly a number of free-living eelworms. If the tulip root is due to *Tylenchus* we shall find a considerable number of *Tylenchus* together with a number of free-living forms. We are now in a position to distinguish *Tylenchus* from the free-living forms (see Fig. 51). In looking for eelworms we simply tease a small portion of the stem in water on a slide and examine it under the microscope.

A *Tylenchus* very similar to *T. dipsaci* in appearance is one of the causes of "clover sickness" and is held by many authorities to be the same as that which attacks oats: nevertheless healthy crops of oats have been grown after clover attacked by the eelworm *Tylenchus* sp.

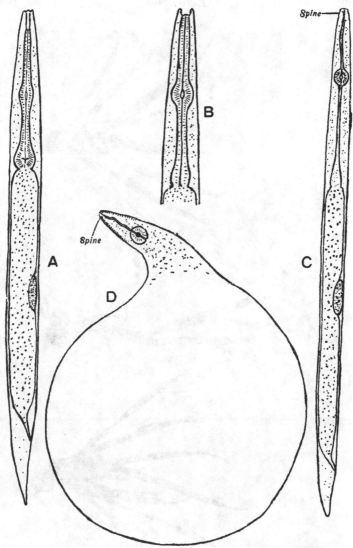

Fig. 51   *A* and *B*, free-living eelworms; *C*, a parasitic eelworm;
D, the swollen female of *Heterodera radicicola*. (Magnified.)

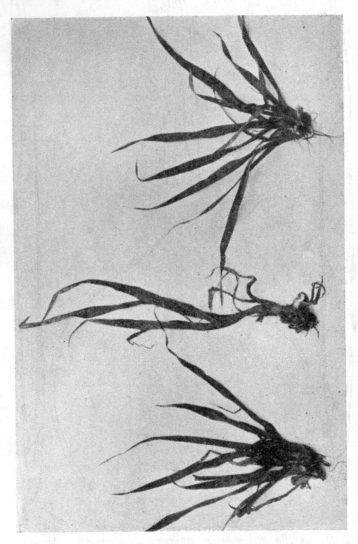

Fig 52.   Oat plants attacked by *Tylenchus dipsaci*.   The twisted stems are very characteristic.

Remedial Measures.

Wheat and barley grow well on infested soil. Red clover can be replaced by sainfoin, trefoil or white clover. Such crops as red clover, oats, buckwheat and bulbs should not be grown for some years on land which is infected.

Fig. 53. Ear cockles of wheat caused by *Tylenchus scandens*. The dark grains somewhat resemble bunted grains.

Ear Cockles in Wheat.

Another disease caused by an eelworm is that known as "purples" or "ear cockles" (Fig. 53).

The culprit in this case is known as *Tylenchus scandens*.

In this disease the grains of wheat are affected. They become rounded or irregular in shape and are of a dark or purplish colour. If a portion of the inside of a diseased grain is examined in a drop of water it is found that the thread-like bodies composing the mass begin to move about, and by their structure show themselves to be parasitic eelworms. The eelworms are said to crawl up on the outside of young plants and enter the developing flowers.

Remedial Measures.

The grains containing the eelworms are very light and float to the surface when wheat is steeped and are easily skimmed off.

Root Knot.

The eelworm which produces knots on the roots of a large number of plants is known as *Heterodera radicicola*. It may also be found at the base of the main stem which is below the ground (see Fig. 54).

This eelworm is peculiar in that the female after being fertilized swells and becomes almost spherical (see Fig. 51 *D*)

Inside the knots formed by this eelworm are small white balls which when examined under the microscope show themselves to be eelworms. The males are very like *Tylenchus*. These eelworms flourish best on sandy soils.

Remedial Measures.

In fields this disease can be reduced by means of gas lime. In greenhouses it may be kept down by watering heavily or by partial sterilization.

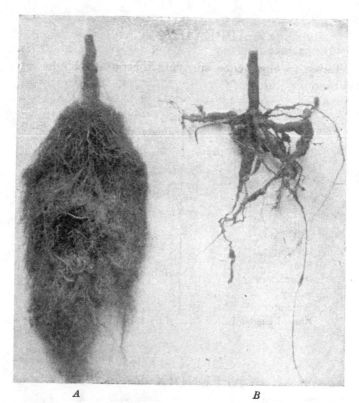

Fig. 54. *A*, a healthy tomato root; *B*, a tomato root attacked
by *Heterodera radicicola.*

The beet eelworm (*Heterodera schachtii*) does not
form knots in the same way, as the body of the female
is outside the plants, only its head being inside.

APPENDIX A

Dates to commence spraying for potato blight in main-crop potatoes.

Cornwall, Devon ... ... Dorset, Isle of Wight and Hampshire ... ... Somerset, S.W. Wales ...	June 15th–end of June
Glamorgan ... ... Gloucester ... ... N.W. Wales ... ... Sussex ... ... Wiltshire	July 1st–July 8th
Berkshire ... ... Hereford ... ... Kent ... ... Oxfordshire ... ... Surrey ... ... Worcester	July 8th–July 15th
Rest of England	July 15th–July 31st

APPENDIX B

The following mixture has recently been shown by the Lea Valley Research Station, Waltham Cross, to be effective in controlling "damping off."

2 ozs. of Copper Sulphate
11 ozs. of Ammonium Carbonate } Powder and mix.

The dry mixture should be kept in a corked vessel 24 hours before using.

1 oz. of the mixture is dissolved in 2 gallons of water.

Vessels of iron, tin or zinc must not be used.

APPENDIX C

WHEAT BULB FLY

(*LEPTOHYLEMYIA COARCTATA*)

An attacked plant can bo distinguished from one attacked by frit fly by the maggot which can be found feeding on the central shoot. This is whiter, has larger jaws (i.e. it looks blacker at the head end) and it has tubercles on the tail end which closely resemble those of the cabbage fly maggot (see Fig. 43 *A*).

The flies hatch out in June and July and lay their eggs *in bare soil* about $\frac{1}{8}$ inch below the surface in July, August and possibly September. Most of the eggs hatch out the following spring, and are usually found attacking wheat plants in March and April. The larva on hatching makes its way into the middle of the wheat shoot where it feeds at the base of the shoot which it kills. When fully fed the larvae make their way into the soil where they pupate about $1\frac{1}{2}$–2 inches below the surface, usually in May.

Remedial Measures.

Wheat on a firm tilth suffers less than on a loose tilth as the plants tiller better and a smaller percentage are killed entirely.

Where wheat is following a bare fallow, if the fallow is clean by the middle of July, then the sowing of mustard for ploughing in later would probably prevent attacks, as the fly only lays its eggs in bare land.

Fig. 55. Fruiting bodies of the clover stem rot fungus as
found in an old clover ley in November.

INDEX

Abdomen 85
Acid manures 43
Aecidiospores 68
Agar-agar 7
Agrotis 98
American gooseberry mildew 14, 55
Amoeba 38
Anbury 35–43
Antennae 85
Aphides 158
Ascospore 52, 59
Ascus 52, 59

Barberry 68
Barley ergot 61
  frit fly on 127
  gout fly on 134
  Hessian fly on 138
  mildew 47
  rust 64
  smut 73
Bean aphis 163
  beetle 126
  weevil 125
Beet rust 1, 72
Beetles 92, 103–127
Birds 90, 101, 118, 148
Black scab of potatoes 44–47
Blight 158
  of potatoes 17
Blue stone 26–30, 78
Bordeaux mixture 26–30, 35, 55
Bordeaux mixture (dry) 29–30
Botrytis 5
Bots 149, 153
Bruchus 126
Bunt 81–84
Burgundy mixture 29
Butterflies 87–92, 102

Cabbage butterflies 85–91, 98
  caterpillars 85–91, 98
  club 35, 122
  finger and toe 35
  gall weevil 122
  moth 98
  root fly 140
  root maggot 140
Calandra 119
Carbon disulphide 121, 127
Carrot fly 145
Caterpillars 87–92, 98
Caterpillars' eggs 90
Cecidomyia destructor 138
  tritici 149
Cells 4
Centipedes 107
Cetonia 113
Ceutorrhyncus 122
Chafers 113
Charlock 36
Chitin 86
Chlorops 134
Chrysalis 89
Claviceps 56–61
Click beetle 103
Clover sickness 62, 166
Clover stem-rot 62
Club 35–43, 122–124, 170
Cocoon 92
Conidia 12
Conidiophore 19
Corn sawfly 164
Crane flies 145
Cultures 7

Daddy longlegs 145
Damping off 31–33 and Appendix B
Dry spraying 29–30

Ear cockles 169

Eelworms 165–171
Epidemic 14
Ergot 56–61
Erysiphe 47–54

Finger and toe 35–43
Flies 92, 127–158
Formaldehyde 78–79
Frit fly 127
Fungicides 26, 55, 78, 79

Gastrophilus 153
Gelatin 7
Gid 155
Gout fly 134
Grain weevil 119
Green fly 158
Gulls 101, 108

Haustoria 13
Hepialus 102
Hessian fly 138
Heterodera 170
Hop aphis 163
    mildew 54–55
Horse bot 153
Hover fly 161
Hylemyia 140
Hypha 3
Hypoderma 149

Ichneumons 90, 162
Imago 89–92
Insecticides 97, 101, 121, 163
Insects characteristics 85
    useful 90, 161, 162

Lady-birds 161
Larks 112
Larvae 89–92
Leather jackets 145
Lime 26–31, 43
Lucilia 156

Maggot 92
Mamestra brassicae 102
Mangold rust 1, 72
    surface caterpillar 98
May bug 113
Melolontha 113
Metamorphosis 91
Midge 149
Mildews 47–56

Millipedes 107
Mole 107, 118, 148
Moths 92, 98–102
Mould 11
Mushroom 8, 13
Mycelium 3
Myxamoeba 38

Nectria 16
Niggers 161

Oats, eelworm in 166
    frit fly 127
    mildew 47
    rust 64
    smut 73
    tulip-rooted 166
Oestrus 155
Onion fly 145
    mildew 33–35
Oscinus 127
Owls 118

Parasite 2
Pea beetle 126
    weevil 125
Penicillium 5, 6
Peronospora schleideni 33–35
Petri dish 7, 8
Pheasants 148
Phorbia 140
Phyllopertha 113
Phytophthora 17–31
Pieris 85–91
Pigs 101, 118
Plasmodiophora 35–43
Plasmodium 39
Plovers 101, 118, 148
Plutella 102
Potato black scab of 44–47
    canker 44–47
    disease 17–31 and Appendix A
    dry rot 17–31
    tumour 44, 47
    wet rot of 24
Preservatives 12
Pupa 89, 92
    cases 92
Puparia 92
Pythium 31–33

Rhizotrogus 113

Rooks 101, 118, 148
Root flies 140
    knot 170
    maggot 140
Rose chafer 113
    mildew 55
Rotation of crops 15
Rusts 64–72

Saprophyte 2
Sawflies 92, 164
Sclerotinia 62–64
Sclerotium 56
Sheep maggot 156
    nasal fly 155
Sitones 125, 126
Smuts 73–84
Spiracles 86
Spore-cases 51
Spraying 26–30
Starlings 101, 118, 148
Sterilization 11, 12, 33
Suckers 13
Surface caterpillars 98
Swift moth 102
Synchytrium 44–47

Teleutospore 67
Thorax 85
Tilletia 81–84
Tipula 145–148
Toads 101
Trachea 86
Tulip-rooted oats 166
Turnip club 35, 122
    finger and toe 35
    flea beetle 107

Turnip fly 107
    gall weevil 122
    moth 98
    root maggot 140
    surface caterpillars 98
Tylenchus 166, 169, 170

Uredineae 64
Uredospore 65
Uromyces 1, 72
Ustilago 73–81

Warbles 149
Ward's Tube 57
Wart disease of potatoes 44–47
Wasps 90, 148
Weevils 119
Wheat bulb fly 140 and Appendix C
    bunt 81
    ear cockles 169
    ergot 61
    frit fly on 127
    gout fly on 134
    Hessian fly on 138
    Little Joss 71, 72
    midge 149
    mildew 47
    rust 64
    smut 73
White grubs 113
Winter moth 94
Wireworm 103
Woolly aphis 163
Wound parasites 16

Zoospore 21